AVERROES' QUESTIONS IN PHYSICS

The New Synthese Historical Library
Texts and Studies in the History of Philosophy

VOLUME 39

The titles published in this series are listed at the end of this volume.

AVERROES' QUESTIONS IN PHYSICS

From the unpublished
Sêfer ha-derûšîm ha-tib'îyîm

Translated and edited by

HELEN TUNIK GOLDSTEIN

School of Religion, University of Iowa, U.S.A.

KLUWER ACADEMIC PUBLISHERS
DORDRECHT / BOSTON / LONDON

Library of Congress Cataloging-in-Publication Data

Averroës, 1126-1198.
Averroes' questions in physics : from the unpublished Sêfer ha-derûšîm ha-tib'iyîm / translated and edited by Helen Tunik Goldstein.
p. cm. -- (The New synthese historical library ; 39)
Includes bibliographical references and indexes.
ISBN-13: 978-94-010-7468-1
1. Physics--Early works to 1800. 2. Physics--History.
3. Science, Ancient. 4. Philosophy, Islamic--Early works to 1800.
5. Aristotle. Physics. I. Goldstein, Helen Tunik, 1930- .
II. Title. III. Series.
QC7.A85 1990
530--dc20 90-47754

ISBN-13: 978-94-010-7468-1 e-ISBN-13: 978-94-009-2151-1
DOI: 10.1007/978-94-009-2151-1

Published by Kluwer Academic Publishers,
P.O. Box 17, 3300 AA Dordrecht, The Netherlands.

Kluwer Academic Publishers incorporates
the publishing programmes of
D. Reidel, Martinus Nijhoff, Dr W. Junk and MTP Press.

Sold and distributed in the U.S.A. and Canada
by Kluwer Academic Publishers,
101 Philip Drive, Norwell, MA 02061, U.S.A.

In all other countries, sold and distributed
by Kluwer Academic Publishers Group,
P.O. Box 322, 3300 AH Dordrecht, The Netherlands.

Printed on acid-free paper

TABLE OF CONTENTS

INFORMATION FOR THE READER

The following abbreviations are used for the titles of Averroes' works:

DSO = *De Substantia Orbis*
EDAn = *Epitome of De Anima*
EDC = *Epitome of De Caelo*
EDGA = *Epitome of De Generatione Animalium*
EDGC = *Epitome of De Generatione et Corruptione*
EM = *Epitome of the Metaphysics*
EP = *Epitome of the Physics*
FM = *Faṣl al-Maqâl*
ICat = *Intermediate Commentary on Categories*
IDC = *Intermediate Commentary on De Caelo*
IDGC = *Intermediate Commentary on De Generatione et Corruptione*
IP = *Intermediate Commentary on Physics*
IPrA = *Intermediate Commentary on Prior Analytics*
ITp = *Intermediate Commentary on Topics*
Kašf = *Kitâb al-Kašf 'an manâhij al-adillah*
KMH = *Kôl Meleket Higgâyôn*
LDAn = *Long Commentary on De Anima*
LDC = *Long Commentary on De Caelo*
LM = *Long Commentary on Metaphysics*
LP = *Long Commentary on the Physics*
LPsA = *Long Commentary on Posterior Analytics*
QLog = *Quaesita in libros logicae*
TAT = *Tahâfût at-Tahâfût*

Other titles abbreviated:

AHDLMA = *Archives d'histoire doctrinale et litteraire du Moyen Âge*
CAG = *Commentaria in Aristotelem Graeca*
CCAA = *Corpus Commentariorum Averrois in Aristotelem*
HU = M. Steinschneider, *Die Hebräischen Übersetzungen des Mittelalters*
JAOS = *Journal of the American Oriental Society*
JNES = *The Journal of Near Eastern Studies*

The Hebrew texts of the Intermediate Commentaries on *Physics* and *De Caelo* are cited from Hebrew MS 42 in the collection of the Harvard library. All references to the Latin translations of Averroes are cited from *Aristotelis omnia quae extant Opera, et Averrois Cordubensis in ea opera omnes, qui ad haec usque tempora pervenere commentarii* (Venetiis apud Iunctas, 1562-1574), unless otherwise noted. Where the edition of B.H. Zedler contains variant readings of the text of *TAT* (*Destructio Destructionum*), reference is made to that edition.

In the notes, quotations from the text of Aristotle are cited in the translation of *The Works of Aristotle* (translated into English under the editorship of W.D. Ross [Oxford, 1928]), unless otherwise indicated.

Quotations from the text of *TAT* are taken from the translation of S. van den Bergh, *Averroes' Tahafut al-Tahafut* (Oxford, 1954), and quotations from *FM* are taken from the translation of G. Hourani, *Averroes on the Harmony of Religion and Philosophy* (London, 1961). Where an English translation of an Averroean commentary is available in CCAA, citations are usually given in that translation. On occasion, I have preferred to use my own, and this is indicated. All other translations are mine--again, unless otherwise indicated.

Readers should be aware that an English translation of the Long Commentary on *Metaphysics*, Book XII (*Lam*) is now available: C. Genequand, *Ibn Rushd's Metaphysics* (Leiden, 1984). Because this translation cites the number of the Arabic page, I have not included the page numbers of the translation in my notes. In the few cases where I have given an English translation of a passage from that work, the translation is mine.

Citations from *TAT* refer first to the English translation of S. van Bergh, *Averroes' Tahafut al-Tahafut*. In other citations from Averroes' writings where I have been able to use a Hebrew text, the citation from the Hebrew text is given first. In the case of *Kašf*, citations not otherwise identified are taken from the Cairo (1935) edition.

Due to the limitations of my software, all long vowels in transliterated words are indicated by circumflex accents. Transliteration of Hebrew ignores the absence of *dâgêš lene*. I have tried to transliterate on the basis of grammatical form, rather than mediaeval orthography. Unfortunately, there is still no generally accepted system of transliterating Hebrew, so I must ask the reader's indulgence for what may seem overly-pedantic or inadequate. There are a few cases in which I could not bring myself to depart from a common, if none too correct, transliteration of a Hebrew or Arabic term or name, but transliteration of Arabic generally follows the system of *The Encyclopedia of Islam*. Long vowels are not indicated in proper names.

In the translation of the *Questions*, material within square brackets is explanatory or specificatory. Pointed brackets enclose emendations. In *Questions VI* and *VII*, material within curly brackets ({ }) is supplied from the Arabic text.

While I hope that this volume, in itself, is readable for all students of mediaeval philosophy, it is meant to be read in conjunction with my edition of the Hebrew text, soon to be published as part of the *Corpus Averrois* by the Israel Academy of Sciences and Humanities. Readers will find that the numbered paragraphs of the translation correspond exactly with the numbered paragraphs of the Hebrew text.

The material in the notes is designed to facilitate the reading of this text. It was not my aim to lay out in great detail all of the philosophic problems to which Averroes alludes. I have tried to cite the relevant passages in Aristotle and the commentators on which Averroes builds. The extensive citation of parallel passages from other works of Averroes should not be interpreted as a desire to impose a false consistency on the author. Frequently, the passages cited reveal the wide range of views it was possible for Averroes to hold on one and the same topic.

AVERROES' QUESTIONS IN PHYSICS

HELEN TUNIK GOLDSTEIN
School of Religion
University of Iowa
Iowa City, Iowa 52242
U. S. A.

INTRODUCTION

1. The Commentator

In the transmission of the Peripatetic tradition, no mediaeval commentator was more influential than the Spanish Muslim philosopher Ibn Rushd (1126-1198), or, as he is more commonly known in the West, Averroes. Just as Aristotle was *the* philosopher to the mediaevals, so Averroes was *the* commentator--a view held by Renaissance philosophers as well. It was the great Renaissance editions of Aristotle in Latin, accompanied by the commentaries of Averroes, that assured Ibn Rushd's place as the authoritative commentator for all time.[1]

In part, Averroes' status was due to the thoroughness of his work: very little of the Aristotelian corpus escaped his pen. For much of this corpus, he offered readers not one, but three commentaries: an epitome (commentary by way of condensation); an intermediate or middle commentary (commentary by way of paraphrase); and a long commentary, where the lemmas from the Aristotelian text are explicated in great detail.[2] He was at some pains to search out the best of the Arabic translations of Aristotle,[3] and he was well acquainted with the work of his predecessors, whether Muslim, or Christian (if their work was written in Arabic), or Greek-- primarily, Alexander of Aphrodisias and Themistius. Since Averroes knew no Greek, he was dependent on translations for both his knoweledge of Aristotle and his knowledge of Aristotle's Greek commentators.[4]

In addition to the commentaries, Averroes authored a number of independent works of a philosophical or theological nature. Of these, the best known is the *Tahâfut al-Tahâfut* (*Refutation of the Refutation*), in which his response to al-Ghazali's attack on the philosophy of Avicenna forms the platform on which Averroes builds the exposition of his own system. Another polemic/apologetic work is the *Faṣl al-Maqâl* (*The Decisive Treatise on the Harmony of Religion and Philosophy*) in which Averroes takes on the Mutakallimûn.[5] The first of these works was translated into both Hebrew and Latin, and therefore had wide influence. The second was never translated into Latin and, therefore, remained virtually unknown until the end of the nineteenth century.[6]

A similar situation exists with regard to the independent works of a more purely philosophical nature. The *Questions in Logic* and the collection known as *De Substantia Orbis* were both translated into Latin (and both printed in the Renaissance editions), while the *Questions in Physics* (*Sêfer ha-derûšîm ha-ṭib'îyîm*) was translated only into Hebrew, and thus did not enter the tradition of Christian scholasticism or become known to the Renaissance.[7]

2. The Text and its Versions

As with the *Questions in Logic*, the origin of this collection, as a collection, is obscure.[8] Although the lists of Averroes' works in the Arabic bibliographic treatises contain references to collections of works on philosophical topics (and perhaps a specific reference to a collection of treatises on physics with the title *Questions* [*masâ'il* = *derûšîm*]), the scope of these collections is unknown.[9] Over half the treatises in the *Questions in Physics* may, with a high degree of certainty, be identified with treatises mentioned in the Arabic lists, but in those lists not all the treatises are called "questions." The terms "treatise" [*maqâla*], "discourse on" [*kalâm 'alâ*] and "appendix" [*ta'lîq*] are also used. It is unclear whether this indicates that the treatises so designated were not included in the collection termed *Questions*.

Only two of the *Questions in Physics* (numbers VI and VII in this edition) are still extant in Arabic, and they exist only in a unique MS. Nothing in that MS indicates that the treatises formed part of a larger collection. The Hebrew text, too, gives us little, if any, information about the existence of an Arabic collection, for the Hebrew text exists in more than one version.

In addition to the version of nine treatises, titled *The Book of Questions in Physics*, where works by other authors may either interrupt the sequence or be added at the end, and where the commentary of Narboni follows each *Question*,[10] we have also:

(1) A version, with the same title, and with the commentary of Narboni, but with only five treatises which are not consecutive, but interspersed with treatises on related topics written by other authorities.[11] Averroes' treatises, however, are found in the same order as the "standard" version, and the translation is also that of the "standard" version.

(2) A version entitled *Questions on Physics and Divine Matters [metaphysics]*, which lacks the commentary of Narboni, and which contains only four treatises, in a different order, interspersed with works of other authors. (This version also contains what purports to be an independent work of Averroes, but is actually an excerpt from Long Metaphysics XII, Comm. 18.[12]) The translation contains certain verbal variants from the "standard" version.[13]

(3) A version, containing only one treatise, lacking both an

overall title and the commentary of Narboni, but in which the treatise is given a close association with *De Substantia Orbis* VII, which immediately follows it in the text.[14]

This third version is the sole case in which a Hebrew translator can be named: the translation was made by Ṭodros Ṭodrosi in the year 1340. The only conclusion to be drawn from his translation is that Ṭodrosi may definitively be eliminated as the translator of any of the other versions. However, we may be able to draw a tentative conclusion as to the formation of the Hebrew collection.

The earliest evidence for the existence of the nine treatise collection is the commentary of Narboni, completed in 1349. The fact that nine years earlier one treatise could be attached to a work outside the corpus may indicate that the Hebrew collection of nine treatises was formed during those nine years, or may even indicate that Narboni himself collected the various treatises.[15]

Narboni, however, was not the translator of these works. In fact, no definitive indication of the translator's identity exists.[16]

3. The Nature of the Question-Form

Steinschneider offered the following general characterization of Averroes' *Quaestiones*:

> These are mostly brief discussions, more or less answers to questions; they may be partially occasioned by topics in his commentaries and may be considered as appendices to them.[17]

In the case of the *Questions in Physics*, this description must be supplemented. At least two of our treatises are expository in nature.[18] Furthermore, of those that do properly belong to the *Aporienliteratur*, the "question" involved must not be taken to be one actually asked of Averroes. Most of the difficulties he deals with had already been discussed, at some length, by other commentators. In terms of form, Averroes is simply following the teaching method of the Hellenistic schools, a firmly established tradition by his time.[19] The Hebrew and Arabic terms which we translate by the word "question" (Latin: *quaestio*), could just as well be translated by the word "investigation."

Again, our collection is not merely composed of works that have a close relation to Averroes' commentaries. A number of them are more closely related to independent works such as *Tahâfut al-Tahâfut*, *Faṣl al-Maqâl*, *Kitâb al-Kašf*, *De Substantia Orbis*,[20] and even the *Questions in Logic*.[21] It also cannot be asserted without qualification that all of these works are appendices to the commentaries (or larger independent works).[22] Some of the *Questions* may be preliminary sketches for Averroes' treatment of a problem in a larger work.[23] Because only one of the treatises is dated,[24] and because it is not always possible to assign a precise date to the larger works, it is very difficult to make statements concerning the relation between works with any degree of certainty.

It is true, as Steinschneider noted, that fewer of these *Questions* have a direct relation with specific loci in the Aristotelian corpus than do the *Questions in Logic* (although at least four of our treatises do have such a direct relation).[25] It is also true that none of the individual treatises in the collection have titles. But this statement is true only of the Hebrew text in the anonymous translation. In the Arabic text, *Question* VII has a title, and *Question* IX, in the translation of Ṭodrosi, also has a title.[26]

While the *Questions in Logic* are all related to Averroes' commentaries on the *Organon*, thus providing a reasonable explanation of the name, not all the *Questions in Physics* are related to Averroes' commentaries on the *Physics*. In addition to the relationship with independent works already cited, *Question* VI is directly related to *De Generatione Animalium* and Averroes' commentary on that work, and also to the Long Commentary on *Metaphysics*. Both *Questions* II and IX are closely related to *De Caelo* and Averroes' commentaries thereon. The one unifying factor in the collection that would justify its title, is that all of the treatises, in one way or another, deal with motion and/or time.[27]

4. MSS and Other Textual Material

The translation of this work is based on an edition compiled from eight Hebrew MSS, the single Arabic MS, and certain other materials discussed below.[28]

Only two of the eight Hebrew MSS used contain all nine treatises. In both these MSS the individual works are arranged in the same order, and this is the order I have followed in numbering the treatises. The text is in the anonymous translation, and Narboni's commentary follows the text of each treatise.[29] These MSS are:

Vaticanus, Urbinates Ebraici 41.[30] Our text extends from fol. 88 through fol. 115b. The title is given as: "*The Book of Questions in Physics* which the philosopher Ibn Rushd composed and, together with it, *Questions in Physics* belonging to other philosophers." *De Substantia Orbis* is included under the *Questions in Physics*, and follows the treatises of our text. At the conclusion of *De Substantia Orbis*, a colophon gives the date of completion of Narboni's commentary to the *Questions*: February 5, 1349.

Munich Staatsbibliothek 31.[31] Our text extends from fol. 256 through fol. 287b and is immediately followed by *De Substantia Orbis*. The title is given as: "*The Book of Questions in Physics* which the philosopher Abu Alid [*sic*] Walid ben Rushd composed, and together with it *Questions in Physics* belonging to other philosophers." There is no colophon.

This MS is very carelessly written and very hard to read, although the scribe has provided artistically written headings and someone has provided a drawing of two delightful sphinx figures (fol. 278). The text, however, is full of the most elementary errors in both orthography and grammar. More serious are the errors occasioned by the scribe's attempts to "correct" the text.

The remaining Hebrew MSS are:

New York, Jewish Theological Seminary of America 2311, formerly Steinschneider 6.[32] The text of the *Questions* begins with *Question* II (fol. 1b), but all the remaining *Questions* are present in the usual order,and in the anonymous translation; *De Substantia Orbis* begins at fol. 49. The opening leaves of this MS are badly damaged by water,[33] and one leaf is missing in the text of *Question* VIII.[34] The importance of this MS lies in the fact that the scribe was apparently able to correct his own work by reference to a MS other than the one he originally copied.[35] This is particularly evident in the text of *Question* VII, where lacunae existing both in the other Hebrew MSS, and in the body of this MS, are corrected by the scribe in the margins. This MS stands closest to the Arabic and generally contains excellent readings.

Paris, Bibliothèque Nationale 988.[36] Our text begins fol. 87 with Narboni's preface (the only MS I have seen which contains it). The *Questions* proper, again in the anonymous translation, begin at fol. 87b with the title: "*The Book of Questions in Physics*." The second treatise[37] begins fol. 88b, and the third at fol. 89. The fourth treatise (fol. 91b) is followed by two non-Averroean treatises. The fifth of our treatises (the last in this MS) begins fol. 95. Immediately following Narboni's commentary (fol. 97), a colophon gives the name of the scribe: Jacob Alkita.

Munich, Staatsbibliothek 36.[38] This MS, together with MS. Leipzig 40g[39], is part of a distinct family of MSS.[40] Our text begins fol. 219b with the title: "*Questions Concerning Physical and Metaphysical Matters*." This is immmediately followed by *Question* I, *Question* V, *Question* IV (fol. 220),[41] and *Question* II. The MS continues with *De Substantia Orbis*, followed by a work entitled: "*A treatise belonging to the philosopher, the judge Ben Rushd*--actually an excerpt from the *Long Commentary on Metaphysics* XII, Comm. 18.[42] This MS is distinguished not only by its title, and by the fact that it lacks Narboni's commentary, but also by some rearrangement of the text between the *Questions*,[43] certain verbal differences, and some incorporation of glosses.

Paris, Bibliothèque Nationale 989.[44] I have discussed this MS, which contains only *Question* IX, elsewhere.[45] We have here the translation of Ṭodros Ṭodrosi, and in this version *Question* IX has a title.[46] This MS lacks the translator's introduction which may be found in the other two exemplars of this version, but all three MSS contain the interesting colophon discussed in my article. The other MSS of this translation are:

Paris, Bibliothèque Nationale 1023.[47] This MS contains the translator's introduction, written very elegantly in a column parallel to the text of the *Question*. Our text begins, again with title, at fol. 159b, and ends fol. 161b.

London, British Museum Add. 27559.[48] Here, too, the translator's introduction is written in a column parallel with the text, but the hand is not so elegant. The text begins at fol. 304, with title, and ends fol. 306b. This MS is a copy of P. 1023.

The first four manuscripts listed all belong to the same tradition of translation, that of the anonymous translation. I do not find it possible to identify any one of them as a prototype; rather, they all appear to depend on a common ancestor. MS. Mun. 36 presents certain problems. Despite the verbal differences already noted, this manuscript, too, is basically a part of the anonymous translation tradition. Its peculiarities are most likely due not to a difference in prototype, but to a different history of transmission. It should be noted that MS. Mun. 31 does exhibits a peculiarity which might indicate a different prototype: the spelling of the name of John Philoponus.[49]

The three MSS of Ṭodros Ṭodrosi's translation exhibit an entirely different style. There are certain indications that this translation may go back to an Arabic MS differing slightly from the Arabic prototype of the anonymous translation.[50]

The sole extant Arabic MS is:

Madrid, Escurial 632.[51] This MS contains only *Question* VI and *Question* VII of our text, and it contains them in reverse order.[52] *Question* VI (fol. 85) is untitled, but *Question* VII (fol. 76b) has the title: "A treatise belonging to Abu'l Walîd concerning the seventh book, and the eighth of [Aristotle's *Physikê*] *Akroasis*" (bracketed words in margin). The MS is written in a difficult North African hand.[53] Derenbourg states that each leaf contains 15 lines, but there are actually 17 lines to the page. Vowel marks, when present, are written under the diacritical points, and hamza is written very rarely. The only form of punctuation is "thick writing," and that is used very sparingly. However, before quotations (both direct and indirect) one occasionally sees an elongated form of the word. The vocalization appears casual and haphazard, but the scribe did engage in some proof-reading: lacunae are supplied in the margins, along with a few suggestions for corrections. The MS is dated Almeria, 724/1324. But following *Question* VII it is stated that the composition of the treatise was completed in Seville, month of the Hadj, 572/1175-6.

It is probable that this MS is not the exemplar on which the Hebrew translation is based. There are a number of differences between the Hebrew version and this MS, not all of which can be explained by assuming a misreading on the part of the translator, or by reference to the most common scribal errors. Furthermore, in the text of *Question* VI, the Arabic contains a number of, for the most part, explanatory phrases which do not exist in our Hebrew version, and which may be glosses incorporated in the text. I believe that the ancestor of the Hebrew version frequently had better readings.[54] On the whole, however, the Hebrew offers a faithful and literal rendering of the Arabic. While the translator does not, for example, render the difference in demonstrative pronouns, or reflect the variety of Arabic negatives, there are only a small number of cases where he has misread the text, and only this affects the sense of the text.

In addition to the MSS, I have collated an edition of the text of

Question III contained in

M. Worms, *Die Lehre von der Anfanglosigkeit der Welt bei den mittelalterlichen arabischen Philosophen des Orients und ihre Bekmpfung durch die arabischen Theologen (Mutakallimûn)*, *Beiträge zur Geschichte des Philosophie des Mittelalters* (Münster, 1900), III, iv. Worms edited the text on the basis of three MSS, all of which have been available to me. I do not always agree with his reading of the MSS, but I have followed his suggestions on certain emendations.

Because it is difficult to separate text and commentary, I chose not to collate the lengthy quotations from the *Questions* contained in works by Narboni and Abravanel. Where appropriate, these quotations may be found in the notes. I also did not include the readings of Narboni's lemmas in his commentary in the apparatus. The lemmas rarely shed light on the text. The exceptional cases are cited in the notes.

5. Methods of Establishing the Text

Since no one of the MSS, including the Arabic, is in itself satisfactory enough to serve as a standard by which the others may be measured, I have not hesitated to choose the best reading from whatever MS it may come. There are, however, two exceptions: (1) in the case of *Questions* VI and VII, it is the Hebrew text that is edited. Although the Arabic text has been used to settle doubtful readings in the Hebrew, and to fill in lacunae, I have not attempted to make the Hebrew text conform to the Arabic text. However, most of the suggested emendations do stem from the Arabic. (2) In the case of *Question* IX, the translation of Ṭodros Ṭodrosi has been used in a similar way. I have chosen to abide by the readings of the anonymous translation whenever possible.

In general, I have tampered very little with the text. I have not attempted to correct the grammar of the MSS,[55] and I have preferred to retain the readings of the MSS, if any sense can be made of them, even if emendation would lead to a smoother text. The punctuation, however, is all mine.[56]

6. *Testimonies*

I have found no reference in any of Averroes' other works which may be taken to refer unequivocally to any of these *Questions*.[57] What may be more peculiar is that Averroes makes no explicit reference to any of his other writings in these *Questions*. The *Questions* do, however, contain explicit references to the following works of Aristotle: *Physics*,[58] *De Caelo*, *De Generatione et Corruptione*, and *Prior Analytics*. Additionally, Averroes draws on material contained in *Metaphysics*, *De Anima*, *Categories*, *Posterior Analytics*, *Topics*, *De Interpretatione*, *De Sophisticiis Elenchis*, *De Generatione Animalium*, *De Motu Animalium*, and *Meteorologica*.

Other authors mentioned by name in this text are: Plato (in *Questions* I, IV, VII), Alexander of Aphrodisias (IX), John Philoponus (VII), al-Farabi (with specific reference to his book, *On the Mutable Existences*;

VII), al-Ghazali (IV), the Mutakallimûn (III, VII), the Ash'arites (IV), "the physicians" (from the context, Galen; VI), Hippocrates (VII), and Avicenna (VIII, including a reference to his "Oriental Philosophy"; IX).

Over the centuries, scholars have, for the most part, ignored these *Questions*. In modern times only Steinschneider and Rosenberg have offered a thorough description of the *Questions*.[59] As we have noted, only *Question* III has been edited previously. *Question* III is also the only treatise to have been translated.[60]

Steinschneider did make use of the text of *Question* VII in his monograph on al-Farabi.[61] It is a major source for the well-known appendix on John Philoponus and the Arabs. J.L. Teicher made use of Worms' edition of *Question* III in his "Avicenna's Place in Arabic Philosophy."[62] H.A. Wolfson translated a short excerpt from *Question* V (as quoted by Narboni).[63] H. Davidson cited *Question* IX in his work on Abraham Shalom[64], and more recently has cited the same treatise when considering Averroes' critique of Avicenna's proof for the existence of God.[65] Davidson also used *Question* VII in an article on John Philoponus.[66] Recently, D. Gutas has quoted from *Question* VIII in his treatment of Avicenna's "Oriental Philosophy."[67]

But it must be admitted that the *Questions* were not much more popular among mediaeval and renaissance authors. Both Narboni[68] and Isaac Abravanel(1437-1509)[69] do refer to the collection by name and quote from it. Simon Duran (1361-1444) cites the title of the collection twice in his *Qeshet û-Mâgên*,[70] but the second reference is certainly not to the *Questions*," but to *Kitâb al-Kašf*.[71] Steinschneider thought that there was also a reference to our text in Duran's *Mâgên 'Abôt*, where mention is made of Averroes' "Epistles."[72] From the context in which Duran's citation appears, one might conjecture that he refers to *Question* IX, but the identification is far from certain.[73]

More recently, G. Vajda conjectured that Isaac Albalag (d. 1292) was acquainted with *Question* III,[74] and that that *Question* is a source for Albalag's discussion of perpetual creation. The context in which Albalag discusses this doctrine, however, is not the context of *Question* III.[75] It is more likely that the author relies solely on *Tahâfut al-Tahâfut*.

H. Davidson, in his work on Abraham Shalom (d. 1492), states that Shalom refers to the *Quaestiones in Physica* at *Neveh Shalom* VII, i, 3, 100b.[76] I am unable to find such a reference there; the quotation from Averroes is from *De Substantia Orbis*. As we have seen, the latter work is sometimes subsumed under the general title of *Questions in Physics*. However, Shalom's position on the problem of generation[77] might indicate some acquaintance with *Question* VI--although he could certainly have arrived at that position by using other Averroean texts.

In Latin Jewish texts of the Renaissance, I have found only one use of the *Questions*: in a work by the physician and translator Calo[nymos b. David] Calonymos (fl. first half of the sixteenth century), written for his patron Egidio of Viterbo.[78] The book consists of arguments designed to refute the Averroist position on the eternity of the world, and contains substantial quotations from *Questions* III, VII and IX in Latin translation. The translation is most exact, and it is even possible to identify the Hebrew manuscript from which it was made: Vatican, Urb.

Ebr. 41.[79] I have found no evidence that these quotations were picked up by other authors.

This neglect of the *Questions*, coupled with the fact that the collection was never translated into Latin, might lead one to the conclusion that these works are in some way inferior--not worthy of being noticed. But although some of the treatises are not up to Averroes' usual standard, this is certainly not true of the collection as a whole.

Although there is little that is radically new in these *Questions*,[80] they are not without interest. (What is radically new is very radical, indeed.) The views expressed in this text are often not consistent with Averroes' statements in other works, and sometimes the *Questions* are not consistent with one another. Because of the problems of dating these treatises, it is not always possible to know which statement represents Averroes' final solution of the problems under discussion. But the *Questions* do enable us to see, very clearly, stages in the development of Averroes' thought. In addition, because the *Questions* contain, in a highly compressed form, material drawn from a great variety of sources,[81] they form a useful introduction to the works of Averroes, and, indeed, to some of the most important problems of mediaeval philosophy.

It is tempting to look for an explanation of the very limited interest shown in this collection by mediaeval authors in a statement found in Narboni's preface to his commentary. He tells us that he composed the commentary for his learned colleagues in Perpignan who wished to continue their connection with him even though he now resided elsewhere.[82] Since all of our manuscripts of the collection are in the anonymous translation, and only MSS. Munich 36 and Leipzig 40 lack Narboni's commentary, circulation of this text, at least initially, might have been largely limited to Narboni's disciples.[83]

7. Some of the Philosophic Problems Discussed

Although Averroes touches on many problems common in mediaeval philosophy, and many problems in various texts of Aristotle, the bulk of the material in the *Questions* focuses on problems relating to the eternity of the world, motion, and time. These are the problems discussed in *Questions* III-V, a good part of *Question* VII, and *Question* IX. The arguments presented by Averroes, both in his own name and in the names of his opponents, depend on the interpretation(s) given to Aristotle's views on the infinite, the nature of the "now," and the nature of necessity and possibility. In these treatises, Averroes never discusses purely religious arguments against eternity, nor does he set out to combat the atomic theory of the Kalâm. His interest is confined to the arguments that have a philosophic ancestry, and more specifically, those arguments whose force depends on showing Aristotle to have been inconsistent.

Averroes upholds the view that the cosmos is eternal. An argument against that view, going back to John Philoponus, is that the eternity of the world means that an actual infinite must have been traversed (that is, an actually infinite number of revolutions of the sphere).

Aristotle himself, however, had denied the possibility of traversing an infinite.

Averroes offers many replies to this objection. First, he rejects the linear view of motion and time that the objection implies. The motion of the sphere is circular, and in the circle there is no beginning and no end unless we arbitrarily mark a segment of the circle. Otherwise, the "now" is always both the end of the past and the beginning of the future. For Averroes, all the revolutions of the sphere, in effect, are really one motion. Furthermore, the opponents' argument rests on the assumption that that a specific motion is the end of motion (because what has an end must have a beginning). If that assumption is not granted, the argument fails.[84]

Averroes also sees the objection as resting on another assumption: that in the series of motions, one motion is the cause of the next. However, if the motions are not related causally, an infinite series is possible. If each motion of the sphere is directly caused by the Prime Mover, which has been shown to be immutable and eternal, the series not only may be eternal, it must be eternal. The activity of the Prime Mover and its existence are inseparable.[85]

Another objection against the eternity of the world, again raised by John Philoponus, ran as follows: Aristotle proved that all bodies must have finite force; the sphere is a body; how, then, can it acquire an infinite force from the Prime Mover unless the corruptible can acquire eternity from an eternal being? On Aristotle's own principles, however, this is impossible: Aristotle demonstrated that the eternal can have no potentiality for corruption.

This objection can be understood in two ways: infinite force with respect to the duration of motion, and infinite force with respect to the duration of existence. In the case of motion, one would expect Averroes' response to be that the sphere does not have such an infinite force, but its mover, which does not inhere in the sphere, does. And, in fact, Averroes says just this when he discusses Aristotle's arguments for the self-subsistence of the Prime Mover. But the sphere itself may be regarded as a cause of motion, and Averroes must argue for its infinity with respect to motivity.

Whether dealing with Philoponus' objection as referring to duration of existence or to duration of motivity, Averroes' response rests, basically, on differentiating the sphere from other, sublunar bodies. Unlike sublunar bodies, the sphere is not a compound; it is simple, and its "matter" is something actual, not potential. It possesses no potentiality except the potentiality for its motion in place. Such a simple body, therefore, may be infinite in duration. It may also be infinite in duration with respect to causing motion. But, being a body, it cannot cause motion infinite in intensity or infinite in velocity, for Aristotle had demonstrated that no body, simple or compound, can contain an infinite force of this kind.[86]

Philoponus' objection, as reported by Averroes, also had another element: he spoke of the sphere *acquiring* infinite force from the Prime Mover, and asked if it were possible for the corruptible to acquire eternity, although Aristotle had demonstrated that the eternal can have no potentiality for corruption. The basic issue raised here is one to which Averroes responds in a number of contexts.

Averroes, like Aristotle, identifies the necessary and the eternal. Therefore, any notion of acquiring necessity or acquiring eternity is repugnant to him. He contests Avicenna's view that there are beings which acquire necessity from necessary causes but which are in themselves possible, and he contests the view, which he attributes to Alexander of Aphrodisias, that the heavens acquire eternity from the Prime Mover. What is possible in its own nature remains possible, and what is necessary has no possibility of being otherwise.[87]

In these *Questions*, Averroes is at some pains to establish that "eternal" does not necessarily mean "uncaused." But the nature of the relation between the uncaused eternal and the caused is expressed differently in different contexts. He maintains that "eternity" is an equivocal term as applied both to the world and its cause, although both are eternal in the sense of "everlasting." If the world which is subject to motion and change were created, it would have to be created in its eternal cause. This is impossible, because there is no possibility in the eternal, while creation requires possibility which, in turn, demands something in which the possibility inheres. On the other hand, it has been demonstrated that the existence of motion requires an eternal, immutable, unmoved mover as its cause. Thus, Averroes concludes, both types of existence must be everlasting, separate but eternally conjoined, and one existence must be the cause of the other. The nature of the causation, however is not specified.[88]

Elsewhere in the *Questions*, Averroes denies not only that corporeal substance of the heavens is merely possible, in itself, with respect to existence, but that the eternal motion of the heavens is merely possible in itself. The motion of the heavens, too, is necessary, requiring an eternal mover only because it is an eternal motion and every motion requires a mover.[89] This is radically new, and, to the best of my knowledge, unparalleled in Averroes' writings. Clearly, such a position raises many problems, both philosophic and theological. It makes even more tenuous the relation between God and the world, and it also reduces the importance and power of God--always a sensitive point among Muslim theologians and philosophers. God is no longer the only being necessary of existence. The universe, in and of itself, is also necessary of existence. And beyond that, not even the activity of the world can be said to be possible. It, too, is necessary in itself. In what sense such a necessary activity requires a mover is not spelled out. In spite of these difficulties, one may say that this position does follow from positions already taken by Averroes. If, existence and activity are inseparable in the case of a necessary being known as the Prime Mover, why should not existence and and activity be inseparable in the case of another necessary being, the sphere?

Averroes is also engaged in these *Questions* in defending Aristotle against the charge of use of improper premises. Aristotle himself had found it necessary to respond to such charges.[90] Nonetheless, the charge continued to be made, notably by Galen.[91] The charge is raised with specific reference to *demonstratio per impossibile*, and takes the form of saying that a premise used by Aristotle, while generally true, is not true or possible in the specific case to which Aristotle applies it.

Averroes solves the problem by following the line already laid down by Aristotle, but in greater detail. First he distinguishes between propositions which are false, but possible, and impossibly false propositions and then he considers what type of conclusion is produced by each type of proposition. It is Averroes' contention that an absolutely impossible conclusion cannot be produced by a proposition which is possible, even if the proposition cannot be true in a given case. He denies that the premises in question are impossibly false: in a *demonstratio per impossibile* one may assume a proposition, for example, that is true of bodies, even though the proposition may not be true of a particular body, and no impossible conclusion will result from this assumption.[92]

A treatise dealing with such matters appears inappropriate to a collection of works on physics until one remembers that Aristotle's demonstrations for the existence and incorporeality of the Prime Mover--and Averroes' restatements of them--depend on precisely this type of proof, and on precisely this type of premise. The propositions and proofs used illustratively in this treatise are, indeed, those used for that very demonstration.

But problems of other kinds also arise in connection with motion. One such problem has to do with the Aristotelian principle that all motion involves contact between the mover and what is moved. Averroes does not discuss the case of magnetic attraction in these *Questions*, but he does discuss the case of the apparent self-motion of the elements in *Question* VII. It is also this problem which accounts for the presence of *Question* VI in a collection devoted to physics. In spite of the biological subject matter, the generation of animals, it is a search for a mover able to be in contact with the embryo throughout the process of generation that motivates this treatise. Aristotle had already ruled out several alternatives, and Averroes, by his silence, rules out another: the alternative of Active Intellect as the mover responsible for formation. It might be argued that Averroes' silence on this matter is not significant, that he is merely reporting on what Aristotle had done in *De Generatione Animalium*. But Averroes says explicitly that he will examine the implications of Aristotle's statements, and apparently he does not find Active Intellect, as *dator formarum*, to be one of them. This is another example of the way in which the *Questions* contribute to our knowledge of the progress of Averroes' thought. *Question* VI would appear to belong to an intermediate stage, before the explicit rejection of *dator formarum* found in the *Long Commentary on Metaphysics*, but later than the *Epitome of the Parva Naturalia* or the original version of the *Epitome of Metaphysics*.

8. Acknowledgements

For financial support over many years of study and research, I am indebted to the Littauer Foundation; the American Association of University Women, for a traveling fellowship; the University of Iowa program for Developmental Leaves for faculty; and the Graduate College of the University of Iowa, for expenses connected with the preparation of the manuscript.

I cannot begin to name all the libraries (and librarians) throughout the world that have made my work possible. I must, however, make mention of the great assistance I received from Keith Rageth who heads the Inter-Library Loan service at the University of Iowa.

I would also like to acknowledge the assistance of Weeg Computer Center, and especially the asistance of Elizabeth Curl, in leading me through the maze of software, formatting, and printers.

My colleagues, both in and out of the School of Religion, have been universally supportive and helpful. I am especially grateful to Professors Paul Greenough, who took time from his own research in London to negotiate with the British Library for me, and John Boyle, who, as Director of the School of Religion, supplied me with the computer on which this book was put together. Over the years, a large number of graduate students have helped with gathering bibliographic information and proof reading. I wish to thank them all, as I wish to thank the students whose reactions and questions have forced me to clarify both my writing and my ideas. My husband, Professor Jonathan A. Goldstein, has supplied far more than the customary moral support. His critical reading of the manuscript has saved me from many errors, and he has few peers as proof reader.

It is through the interest and kindness of Professor Norman Kretzmann that my book appears in The New Synthese Historical Library. My editor, Maja S.M. de Keijzer, has been both long-suffering and helpful. No question of mine has ever been too stupid for a thoughtful reply.

My greatest debt is owed to my teachers. No student was ever so blessed in teachers as was I. They truly constituted a generation of giants. Above all, I am indebted to my father, the late Louis J. Tunik, and to the late Professor Harry A. Wolfson, both of whom taught me more than I can say. It is a source of deep regret for me that the two people who would most have enjoyed the appearance of this book are no longer here to see it. This book is dedicated to their memories.

AVERROES' BOOK OF QUESTIONS IN PHYSICS

Question I

The focus of this *Question* is textual and formal. Averroes sees *Physics* VII, 1 as consisting of an introduction followed by a more substantive section. In the introduction, the Arabic translation of *Physics* VII, 1 is not always as explicit as the Greek text, and this forces Averroes to speculate about the precise meaning of Aristotle's text and the purpose for which it may have been written. He also is concerned to clarify the meaning of a technical term (*per se*) as it is used in this chapter. His discussion of the remainder of the text is largely concerned with an analysis and elaboration of Aristotle's argument, which culminates in a reconstruction of that argument in the first figure of the syllogism.

1. Averroes the philosopher said[1]: Aristotle's[2] intention at the beginning[3] of the seventh book of the *Physics* was to demonstrate that everything in motion has a mover.[4]

2. He began by dividing things in motion into those which have an external principle of motion and those whose principle of motion is within them. He assumed [for the purposes of this argument] that the class of things in motion whose principle of motion is within them is moved *per se*, meaning [here] by the term "*per se*" what he [elsewhere] meant by the term "primarily."[5] He explained [elsewhere] what being moved primarily[6] means by saying, "so that it is not in motion because some part of it is in motion." For if we were to consider something to be in motion *per se* simply because it is not moved by an external mover, it would resemble that which is *not* moved *per se*: i.e., that which is moved by part of itself. Thus, this criterion [that is, mere lack of an external mover] could lead us to believe that what is not in motion *per se* is in motion *per se*.

3. Aristotle [here] warned us away from such an inference[7] by saying,[8] "I say first that our supposition that AB is not moved by anything because it is moved in its totality and because its motion is not caused by any external mover is just as though someone should suppose that when DE moves EF and is itself moved, DF[9] is not moved by anything, [simply] on the ground that it is not known[10] which of the two moves the other: whether DE is moved by EF, or EF by DE."

4. But it is possible[11] that Aristotle meant by the words, "I say first that our supposition," up to the end of what we have quoted, that when we see a certain object in motion--AB, for example--which is moved in its entirety without having an external mover, we might suppose that it moves itself and that the mover in it and what is moved are the same thing with no differentiation[12] at all between the two of them.

5. This opinion resembles our own opinion that when, in the same way, we see a certain mover, itself moved, which causes motion in some other thing and which is continuous with that thing and is the source of its motion[13] (for example, DE which moves EF and is itself moved so that DF[14] is moved in its totality), then it appears to the senses that it is

not moved by an external mover [but] that DF moves itself and that it is not moved by any [other] mover at all; and this on the ground that the mover and what is moved by it [appear to the senses to be] one thing with no differentiation at all between them, because it cannot be known [by the senses] which of the two moves the other: whether DE moves EF or EF moves DE.

6. It is as though Aristotle made an allusion in this argument to Plato's opinion that there is something which moves itself and in which the mover and the moved are one [and the same] thing, with no differentiation between them at all.[15] This explanation of the intent of Aristotle's discourse at the opening[16] [of this book] is most plausible.

7. When Aristotle had cautioned us against the aforementioned supposition, he introduced a self-evident proposition, namely, that what is not set in motion by something else will not ever[17] cease to move when something else comes to rest. If this proposition is true, the proposition which results from the conversion of its obverse[18] must also be true, namely, that what comes to rest when some other thing comes to rest must be set in motion by some other thing. With this proposition, which is the conversion of the obverse [of the original, self-evident proposition], taken to be true, he began to demonstrate, by way of a *demonstratio per impossibile*,[19] that everything moved primarily must be moved by something else. [He did this by] assuming an object AB which is moved primarily and dividing it into two parts, for everything in motion is divisible, as was proved in the book preceding the seventh book.[20] He next assumed that one part of this object comes to rest. It follows, then, that because the part is at rest, the remainder of the object must necessarily[21] come to rest, for if the part came to rest while the remainder of the object continued in motion, the whole object would not have been in motion *per se*, i.e., in motion primarily;[22] but it was originally assumed that the whole object was moved primarily.

8. This is an impossible and absurd self-contradiction, and what led to the absurdity must itself be absurd.[23] But this absurdity resulted only from the assumption that part of the object comes to rest while the remainder of it is still in motion. Therefore, the original assumption that, when part of that which is moved primarily comes to rest, the remainder comes to rest, must be true. If this is true, what follows from this assumption must also be true, namely, that what is moved primarily must be set in motion by some other thing,[24] and that when something comes to rest because some part of it comes to rest, it must necessarily be moved by some principle [of motion] in it.[25]

9. The reconstruction of Aristotle's argument in the first figure of the syllogism would be as follows: That which is in motion primarily [and] *per se* is that which comes to rest when some part of it comes to rest. But what comes to rest when some part of it comes to rest is necessarily set in motion by something else. Therefore, what is in motion primarily must necessarily be moved by something else.[26] It has, thus, been demonstrated that whatever is in motion must be set in motion by something else,[27] and this is what we intended to demonstrate.

10. The *Question* is completed. Praise be to God![28]

QUESTION II

The underlying problem addressed in this treatise is to account for the fact that Aristotle proves the perfection and priority of circular motion in both *Physics* VIII and *De Caelo*. Averroes is of the opinion that Aristotle, like God and Nature, does nothing in vain; therefore, it is necessary to find some kind of distinction between the two sets of arguments. The problem, however, is not addressed directly until the very end of the treatise. The remainder of the treatise is devoted to an analysis of Aristotle's arguments, without specification of the source of the arguments. Again, Aristotle's discussions are reduced to syllogistic form.

1. Another treatise of Averroes.[1]

2. I say: The proposition that asserts that circular motion is perfect is the conclusion of a syllogism[2] in which the middle term is the definition of perfection. The syllogism is constructed as follows: The perfect[3] is that which does not admit of increase or diminution;[4] but circular motion [is a motion] which does not admit of increase or diminution:[5] whence it is to be concluded that circular motion is perfect. This [conclusion may be derived] from two affirmative [propositions] in the second figure [of the syllogism] because the major premise, being a definition, is convertible universally.[6] Aristotle frequently employed this kind of syllogism using definitions and properties [as the middle terms].[7]

3. Having established this as true, Aristotle proceeded to conclude, again by means of two propositions, that rectilinear motion is imperfect. [He did this] through the same kind of syllogism as that through which he deduced that circular motion is perfect, namely, through [a syllogism in the second figure composed of two affirmative premises, in this case with] the definition of the imperfect [as the middle term]. Aristotle constructed the syllogism as follows: The imperfect is that which may be increased and diminished;[8] but rectilinear motion is a motion which may be increased and diminished: whence it is to be concluded that rectilinear motion is imperfect.[9]

4. After he had combined the two conclusions, one stating that circular motion is perfect, and the other that rectilinear motion is imperfect, he added to them a third proposition, namely, that the perfect is prior in nature and excellence[10] to the imperfect of the same genus.[11] From these propositions he concluded that circular motion is prior to the imperfect. This kind of syllogism is that which the Peripatetics called "the syllogism brought to a conclusion by means of an additional proposition,"[12] for this syllogism can be brought to a [valid] conclusion only when the following statement is added to it: that both things being in the same genus, one perfect and the other imperfect, the perfect of that genus is prior in excellence to the imperfect by nature, not by position.[13]

5. After the truth of this demonstration had been established, Aristotle saw that he might, perhaps, be asked why circular and rectilinear motion are of this nature [with respect to perfection.][14] He responded that this is due to the interval and magnitude which is the cause of the

specific differentiation of motion, maintaining that circular motion is circular only because it takes place over a circular magnitude, and rectilinear motion is rectilinear only because it takes place over a rectilinear interval.[15] Then Aristotle demonstrated, by propositions analogous to those which he employed with respect to motion, that is, from the definition of the perfect and the imperfect, that the circular interval is by nature more perfect than the rectilinear interval.[16]

6. Once this has been settled it will have become clear, God willing, with respect to which proposition in this discourse doubt occurs, and into which figure of the syllogism our response must fall.[17]

7. You ought to know that of the proofs which exist concerning one and the same thing,[18] there is a proof which functions like evidence and testimony,[19] inasmuch as these establish[20] the form of that thing which is to be demonstrated by proof.[21] There is also a [demonstrative] proof which pertains to the figure and form[22] of that thing, and still another proof which pertains to the perfection of that form.[23] For this reason, the sides which describe the figure of the form are set down first; after this, then, they add the colors which [pertain] to the perfection of the form.[24]

8. With respect to the topic of motion as it is treated in this *Question*, the proofs which pertain to the form are [to be found] in the eighth book of the *Physics*, while those which pertain to the perfection [of the form] are [to be found] in *De Caelo*.[25]

QUESTION III

This is an apologetic treatise, dealing with the sensitive problem of creation. Al-Ghazali had condemned the Peripatetic philosophers of Islam as heretics because of their views on three religious problems: creation, prophecy, and the afterlife. Here, Averroes confines himself to creation and attempts to show that the Mutakallimûn of Islam hold exactly the same principles about what is created as the Peripatetics. The dispute between the parties, according to Averroes, is due to the Mutakallimûn's misuse of those principles, which leads them into self-contradiction, and their misunderstanding of certain basic concepts, such as the nature of time. While no Aristotelian text is under direct examination here, the argumentation on both sides is replete with Aristotelian principles.

1. Still another treatise which also belongs to Averroes.[1]

2. In this treatise we intend to demonstrate that what the Peripatetics believe about how the world came to exist and what the Mutakallimûn of our religion[2] believe about this is of <very nearly> the same significance,[3] most of their disputes being due to the equivocal nature of the terms "eternal" and "created."[4]

3. This is the case because what is described as "created," in the true sense of the term, implies a combination of the following characteristics: that it comes out of something;[5] that its privation precedes it in time;[6] and that it is in time.[7] By the latter I mean to say that

time measures its existence,[8] and contains it, and exceeds it at both extremes [of its existence],[9] and that time is prior to it in nature, and, in sum, is a condition for its existence[10]--although "prior to it in nature"[11] in itself carries this implication: namely, when what is described as "created" exists, or is imagined to exist, time must exist [or be imagined to exist]; and if time does not exist, that described as "created" cannot exist. Still other characteristics [of the truly "created"] are that it has a place prior to it in nature,[12] and an efficient cause which causes its transition from the privation [of existence] to existence and from the possibility of existence to actual existence.[13] What is described as "created" in the true sense of the term, then, possesses all these characteristics.[14]

4. The Mutakallimûn of our religion, however, do not believe that the world comes out of something, or that its privation precedes it in time, or that the world is in place[15] or time. They cannot maintain that it is in time, because time is nothing more than a consequence of the motion of the sphere or, as they say, a concomitant to body.[16] If, therefore, there is no other body before the world, prior in nature to the world, there can be no time at all before the world and time cannot be a condition for the existence of the world or prior to it in nature. The contrary, rather, is the case: namely, the world, that is to say the celestial body, is a condition for the existence of time and prior [in nature] to time. The relationship between the world and place is of a similar nature: namely, the world is prior in nature to place and is a condition for the existence of place, and just as place is confined within the world,[17] time is within its confines.[18] Such being the case, the only characteristic of what we have described as "created" [in the true sense of the term] to be found in the world is that the world exists through [the agency of] an efficient cause.[19]

5. But, in spite of their own postulates, the Mutakallimûn nevertheless consider the world to be finite, and they measure its existence by time,[20] and posit a temporal beginning for its existence.[21] Thus, they maintain that what must be measured by time is the alteration of created things[22] which have a beginning.[23] They attempt to demonstrate this by asserting that what has occurred in the past has already been completed,[24] and what has been completed is necessarily finite and necessarily has a beginning, because what has no beginning cannot be complete. For the Mutakallimûn, therefore, it follows necessarily that time measures the world's existence, is a condition for that existence, is prior to that existence by nature, and includes that existence. Similarly, if they assume the world to be in place, it follows that the relationship between the world and place is of an identical nature. That this is <absurd>[25] may be seen as follows: if the concomitance of time with the body of the world (that is to say, the heavens) measures the world's existence, time must be a condition for the existence of the world; thus it follows that there must be another body before the world which is a condition for the existence of time, and that that body cannot be in time at all. The celestial body, therefore, cannot be *in* time, nor can time measure its existence, since the antecedent is not a condition for the consequent, nor is the consequence [of the condition] necessary for the existence of the condition.[26] Moreover, the relationship between time and the world is of the same nature as the relation-

ship between place and the world: that is to say, the world is a condition for the existence of place, not place a condition for the existence of the world.[27]

6. If time does not measure the existence of the world in its totality,[28] since it is not the nature of the world to be *in* time, but, rather, time exists *when* the world exists, it is clear that the world does not exist in time which measures its existence.[29] Since this is so, it is not true that the world is *in* time, whether that time is finite or infinite.[30]

7. The philosophers say that time has no beginning for the following reason: since time is a consequence of a circle, it follows that it is itself a circle.[31] Just as any point assumed in the circle is both beginning and end to some arc assumed in the circle, so any instant assumed in the circle of time is both beginning and end to some time assumed in that circle, namely, the end of the past and the beginning of the future.[32] For this reason it is true that time, as a whole, has no beginning, just as it is true that the circle has no beginning, inasmuch as time is a circle: for time is confined by a circle and that circle contains it.[33] Consequently, time is not to be assumed to be like the straight line which has no beginning,[34] nor can created things be assumed to be in straight lines which have no beginning; rather, [the process of] creation is circular.[35]

8. If time is imagined to be something abstract,[36] outside of the circle [of the celestial body], it is imagined to be something other than it really is. In accordance with this kind of supposition, it would follow that time is in a straight line and that it has both beginning and end. But when time is imagined to be what it really is, and [imagined to be] a circle sown amongst the spheres,[37] it does not follow that what has gone past of time has come to an end, for if it had an end it would have to have a beginning. Because the Mutakallimûn assume that what occurred [of time] in the past has come to an end, it follows necessarily [for them] that it had a beginning; but because the philosophers do not accept the statement that it has an end, inasmuch as it had no beginning, they do not become involved in the self-contradiction of the Mutakallimûn.[38] The latter became involved in this self-contradiction because they composed a hypothetical syllogism in which the affirmation denies the consequent so that the contradictory of the antecedent is affirmed.[39] If this affirmation, namely, that what has occurred of created things in the past has come to an end, is not conceded to them, since what has no beginning has no end, just as what has no end has no beginning, the syllogism is invalidated.

9. The parts of the motions of the sphere which can be imagined, therefore, have neither beginning nor end, for everything that has an end necessarily has a beginning, but what has no beginning has neither end nor limit. Thus, the assertion of the Mutakallimûn that the created things and the motions which have occurred in the past have actually[40] come to an end is meaningless, for they had no beginning, and what has no beginning has no end.[41] That they had no beginning follows from the demonstration that they are in a circle and from other proofs which they[42] introduce on this point. The totality of created things, therefore, has no beginning although each of its parts is in time, just as each part of the circle is in the circle and has both beginning and end

while the circle in its totality has neither beginning nor end.[43] The totality of created things and the totality of time are of a similar nature.[44] For this reason, one who says that the world is eternal, thinking that the world exists *in* time which has no beginning, errs decisively[45] against the principles of the Peripatetics, in the same way as one who asserts that the world is *in* time which has a beginning errs necessarily.

10. The ancient philosophers who asserted that the world was created agreed that the world is created out of something,[46] but one of them generated time along with the world,[47] while others assumed that time was eternal.

11. The principles of the Mutakallimûn with respect to the creation of the world, therefore, are identical with those of the Peripatetics. They disagree only about what follows from these principles: whether time measures the existence of the world, or does not measure it. But the principles which both schools assume assert that time does not measure the existence of the world, as we have said, and that the world is not to be described as eternal or created because of time. On the contrary, if the world is created in the true sense of the term, it is to be called created only because its existence is forever subject to change. For this reason, the term "created" can be applied in a truer sense to the world, which is in constant change, than to those things which are to be found in change at one time and without change at another, if, indeed, there are beings of this description.[48]

12. Because the great mass of men understand by the term "created" only what is in time and created out of something, they understand the term "created" as applied to the world in the Precious Book by these two characteristics and by forsaking that sense which is the true sense of the term "creation" which is not spoken of [directly in the Quran, and] which must be extracted by means of [the indirect] references pertaining to this in the Book, such as God's dictum, blessed be He!, "I have created the heavens and the earth," and other verses and texts referring to this matter.[49] But in the case of one whose heart does not move him [to believe] that what was created [was created] except from something and in time, it is necessary that the question of that thing out of which God creates be answered in some way;[50] and [if] his intellect cannot be moved from the question of whether or not time was created, something of what the philosophers understand about this matter must be transmitted to him. One who does this, however, has already perverted the scriptural teaching and the prophetic intention.[51] It is for this reason that these things ought not to be set down in a book, but the disciplines peculiar to philosophers ought to be transmitted orally, as is thought to have been the custom of the ancient philosophers who were perfect [in understanding] with respect to both the exoteric and the esoteric.[52]

13. God directs to the truth, and He directs us to the truths we desire to attain, blessed and exalted be He! Amen.

14. The *Question* has been completed. God be praised![53]

QUESTION IV

In this treatise, the argument of Averroes' opponents in *Question* III, the Mutakallimûn, is attributed to Plato. Here, Averroes' refutation of the argument depends not on clarifying the nature of time, but on distinguishing between series in which the members are related essentially and series in which the relation is accidental. Greater stress is placed on the nature and role of the Mover and, although we began with Plato, we conclude with the Ash'arites and al-Ghazali whose conception of the Mover is thought by Averroes to be a gross error.

1. Still another treatise which also belongs to Averroes.[1]

2. Averroes said: The method which Plato followed in order to establish the creation of motion[2] was to say that if a motion is created which had not previously existed, the motions before this motion must either have had a beginning or have had no beginning. If they had no first motion, it follows that these motions which have actually[3] existed could not have existed until infinite motions had been completed before them; but the completion and end[4] of what is infinite is impossible; thus, the existence of this aforementioned motion must be impossible. There can be no doubt, therefore, but what these motions had a first motion, and that the motions before this [last] motion are finite. Consequently, there must have been a first, created motion.[5]

3. Aristotle would have accepted this conclusion with the addition of the following two conditions: one, that the motion which exists in the present is complete and final and that it is the end of the motions, and [two], that one motion is a condition for the existence of another; that is, that the first is a condition for the existence of the second[6] [and so on]. Indeed, Aristotle himself used this argument in the third book of the *Physics*[7] where he said that if the created motion had a motion before it, and that motion had another motion before it, and this proceeded to infinity, the last motion would not exist. He used the same argument in the second book of *De Generatione et Corruptione* when he said that if earth came from water, and water from air, and air from fire, and fire from something else, and this proceeded to infinity, neither the earth nor any other element would exist, since for any one of them to exist it would be necessary for what is infinite to be completed.[8]

4. If, however, these motions are assumed to exist, with no initial motion, because of an eternal prime mover whose action has no beginning (on the ground that what has no beginning to its existence has no beginning to its action),[9] and if one motion is not assumed to be prior to another except by accident, then it is possible for infinite motions, that is, motions with no first motion, to exist before that motion which was posited to be the last.[10] The reason for this is that if we assume motions which have no first motion it follows that there must be an eternal mover whose action has no first action, for it is such a mover that would give completion to the [series of] motions which have no beginning and which cannot come from one another essentially, or have a last motion.[11] Consequently, the supposition[12] [of motions without a first motion] is possible if the motions are related accidentally, but impossible if they are related essentially.[13]

5. Plato, however, in his argument, assumed an essential relationship in place of the accidental, so that his conclusion was true only accidentally. Plato also assumed that the motion which exists at any instant is last, but in the motions which come from one another there is no last (just as in the case of the heat created by some particular motion of the heavens);[14] for if there were a last [motion], no motion could exist after it, in the same way as, if there were a first motion, there would also have to be a last motion. But if there is no last motion in the series, there can be no first; for if, in such a series, there were a first motion, it would follow that one infinite could be greater than another.[15] For this reason, no motion of the heavens can be either first or last. Hence, those who argue after Plato's fashion err in two respects: first, because they assume to be last what is not last, and second, because they assume a relation that is accidental to be essential.[16]

6. The eternity of motion can also be demonstrated from the fact that if the eternal mover who is eternally immutable were at any time in the past without the action of causing motion, then it would be inconsistent with his own nature[17] for him to be the cause of motion [at any later time], just as it would be inconsistent with the nature of one who is eternally immutable to be subject to change, unless the necessary can become possible and what is impossible can become possible.[18]

7. But if Plato's argument[19] is accepted, nothing appears to follow from it except that the first created motion is the first of the motions, [in the sense that there is] no possibility of [the existence of] a motion before this motion [in this world]. Accordingly, it would [still] be possible that there should be infinite worlds prior to this world.[20] Furthermore, if this world were created after it had not previously existed, the possibility of its existence must have been prior to it, and if its possibility of existence had been prior to it, this possibility would have had another possibility prior to it, and the possibilities would have proceeded to infinity; but if a first possibility did not exist, a first actuality could not have existed.[21] This is what one ought to believe[22] according to scriptural teaching,[23] in agreement with the literal[24] sense of the Precious Book ; for, indeed, in various books and compilations of traditions there exists the tradition[25] that there is a world both before and after this world.

8. On the question of the eternity of motion, the Ash'arites are in accord neither with nature nor Scripture,[26] but al-Ghazali[27] thought he had found a decisive argument on this subject, so that he believed that God, blessed be He!, is able to do what seems impossible to our intellect.[28] For he believed that the proof of the Ash'arites is true, namely, that the First, Eternal One has an initial action, and [he also believed] that, with reference to our intellect, God undergoes change by reason of His having actions which have a beginning.[29] But because, according to al-Ghazali, this implies no change in God [Himself], on the ground that His eternal power to perform these actions has no beginning, he asserted that God is able to do what to us seems impossible, and this is the grossest of errors.

9. It is by this kind of consideration that the argument on the existence of motions having no first motion is to be explained,[30] and this is what we intended to prove.

10. The *Question* has been completed.

QUESTION V

While *Question* III discussed the term "created" at great length, this treatise is focused on the term "eternal." It is Averroes' contention that his unnamed opponent (very probably, al-Ghazali) has not understood the equivocal nature of the term, and thus has not realized that to call the world eternal is not to equate the world with God, nor to impugn God's causality. In addition, we have arguments to support the eternity of the world, arguments which turn on the immutability of the eternal.

1. Still another treatise which also belongs to Averroes.[1]
2. Averroes said: Time is a term which signifies the duration[2] of the existence of beings subject to motion,[3] and for this reason we can have no conception of time except along with the conception of motion.[4] Eternity, which is *dahr*[5] in Arabic, is a term signifying the duration of existence of beings which are not subject to motion,[6] and, therefore, it is said that these beings are not in time.[7]
3. Then, if the beings in the existence which is eternal, (that is to say, *al-dahrî*) cannot be in the existence which is subject to motion, the beings subject to motion cannot be in the existence which is eternal (that is, *dahrî*).[8] Thus, the two existences are separate.[9] If this is so, the things which are subject to motion exist only in the existence subject to motion, and, if this is so, both of the existences are eternal.[10] (But even if the existence which is capable of motion is eternal, still it cannot be void of the disposition [for motion], since nothing can be stripped of its nature.)[11]
4. Again, if motion in its totality, and time along with it, were created, its existence, that is to say its creation, would have to be in the eternal existence, since there are only these two existences.[12] Were such a creation possible, it would be possible for the nature of the eternal existence to be converted to the nature of the existence of motion [which is not possible].[13]
5. Consequently, the existence subject to motion must have been eternally conjoined to the eternal existence, and the eternal existence must be a cause for the existence subject to motion,[14] as has been demonstrated in metaphysics.[15] Therefore, if one says that the eternal beings are eternal, and one also says this about the beings capable of motion, the term "eternity," as applied to the two classes, is used equivocally.[16] Taking these things into consideration, one may truly say that the existence subject to motion is an effect of the eternal existence.[17] But because certain men ignored this fact, the two natures were confused. They maintained that if the world were eternal it would have no cause,[18] ignoring the fact that the term "eternal," where applied to both the world and its creator, is used equivocally.[19]
6. In sum, then, the two existences are as different as possible, while their connection is a necessary thing; for if they were not connected in perpetuity (I mean, if one of the two existences existed

first, separately,[20] and then was converted to the second existence), the impossible could become possible. For the existence capable of motion to exist without the eternal existence is clearly absurd, unless the effect can exist without the cause; while from the separate existence of the eternal existence, without the existence subject to motion, it would follow that the existence of the latter is impossible, unless it is possible for the nature of the eternal to be converted to the nature of time and motion.[21]

7. Understand this, for it is the most obvious meaning and the most perfect conception of the two existences. One who does not understand this does not understand motion or time, nor does he understand perfectly either the eternal existence or the second existence. But for one who does understand, all the doubts which occur about the creation of the world and its eternity are resolved, and he also understands the significance of His word, praised be He!, "His throne had stood ere this upon the waters,"[22] that is to say, the connection of the throne and what is below it.

8. I also say[23] that every being which exists in eternity is subject neither to generation nor corruption. For the being in eternal existence (that is to say, *dahr*), is as if it were one [and the same] thing perpetually, and were it corrupted, it would be corrupted at a time in which it [also] exists and, consequently, it would be [both] generated and corrupted at one [and the same] time. But this is impossible. For this reason, what follows is that generation and corruption are truly[24] contraries.[25]

9. And God, blessed be He!, knows the truth.[26]

QUESTION VI

In this treatise, Averroes attempts to determine just what the agent responsible for the generation of the embryo is; how this agent relates to the father; what the nature of this agent is; and the relation of this agent to the faculties of the soul. This determination involves an inquiry into semen and seed, and the embryo, as well as consideration of the instrumentality of this agent, and the nature and instrumentality of the faculties of the soul.

1. Still another treatise which also belongs to Averroes.[1]

2. Averroes said: The intention of this treatise is to investigate the {substance} of the powers existing in the semen[2] [of animals and the seed of plants][3] which resemble[4] that which possesses semen [or seed]; first, in the semen, and second, in the embryo conceived in the womb before its generation is completed. If these powers exist, we shall also investigate what sort of existence they have, and whether they are generated and corrupted or neither generated nor corrupted, as well as the rest of what is desirable to know about them.

3. We say that because Aristotle found there are some bodies which, when they touch another body,[5] bestow on that body the form which is in it and the qualities through which it is what it is;[6] and because it is

evident that the embryo is a body generated by a specific form, namely a form of the {species}[7] of the form of the father[8] {from whom} the semen comes in animals or [the seed] in plants which reproduce by seed: Aristotle began by investigating, first, whether this generator {of the embryo} can be a body; and, if it is a body, whether it is external, like the father, or again, a certain body which separates itself from the father and joins itself to the matter of the embryo immediately,[9] so that it generates[10] the embryo in the womb. [Secondly, he investigated] whether the generator of the embryo can be a power[11] in the semen {which separates itself}[12] from the father and comes to be in the womb and joins itself to the matter of the embryo when the latter is being generated. If the generator of the embryo is this kind of power, what is its substance? What kind of existence does it have, first in the semen, and second in the embryo? Is this power the proximate mover[13] in generation, or does the proximate mover have {another power}?[14] If the latter is the case, what is the substance of this power? Is it in a body or not in a body? Finally, if it is in a body, what sort of existence does it have in that {body}?

4. This, then, is the nature of the questions we must investigate with regard to this matter. These questions [and the answers] exist in Aristotle's words, but some of them are there explicitly, while others are implicit in the principles he posited.[15] We ourselves will begin with what exists explicitly in Aristotle's words, and then [we will take up] what exists implicitly in his principles.

5. We say that the first thing with which Aristotle began was the investigation of the first question: namely, whether it is possible for the generator {and creator} of the embryo to be an external body, e.g., the father, in such a way that that external body is in itself sufficient for this act--even though Aristotle assumed this absolutely [for purposes of argument],[16] in the same way as controversial, ambiguous statements {are assumed}[17] until, when they are qualified with valid conditions,[18] the contradictions are resolved and the nature of the demonstration becomes evident.[19]

6. We say that because Aristotle assumed first that the generator is an external body, that body must inevitably be either {the father} or some other external body besides the father. If we assume it to be the father alone, it follows that the father must be in contact with the embryo until the embryo's generation is completed, for this is the case with bodies that generate one another externally.[20] But the father is separated [from the embryo] and disappears[21] while the embryo continues to be generated,[22] for the {father} is in contact with the matter of the embryo at only one time, namely, at the time of coitus. But if we assume that that which continues[23] the generation of the embryo {after} [the separation of the father] is another external body, this is in contradiction to what we perceive: for we do not find such a body to exist either in the genus of plants generated by seed, or in the genus of animals which procreate by means of semen. Nor can we say that there is a body *in* the womb which produces the embryo, for that body would have to be endowed with soul[24] and life {and would itself} require a begetter.[25] But, again, we do not perceive anything of this description to be begotten from such a thing without semen. {Thus,} it would follow that the generating thing in the semen would itself be produced by something

external that possesses semen, through the medium of semen.[26]

7. We cannot say of the animal that from each member of the begetter there proceeds[27] a similar member [in the semen], for the absurdity of this has already been demonstrated.[28] Moreover, we see that the members of the embryo are generated in succession, and not simultaneously.[29] Furthermore, if the members were generated simultaneously, the embryo would be produced according to latency,[30] i.e., *kumûn* [in Arabic], not according to generation. But the impossibility of all this has already been demonstrated.[31] It would also be impossible for any member to {live}[32] when separated from its associate[33] and then become, of itself, conjoined to it, so that they become one. All this is absurd.[34]

8. In addition, we cannot say that one member in the semen is that which generates the rest of the members, e.g., the heart.[35] Were that the case, the form of all the members would exist actually in the heart, as is the case with the form of what is made [artificially] in relation to the artisan.[36]

9. Then, inasmuch as that which is potential always comes [to be actualized] only from that which is in perfect actuality, both in art and in nature,[37] the only remaining alternative is that there is a procreative power in the semen, for the semen resembles that from which the semen comes: that is, they agree in species {or genus}.[38] Because this power is only potentially the form of that which is generated, not actually; and the potential that exists in art or nature, in every case becomes actualized only by means of what is actual [and] like it,[39] it follows that that which stands in the relation of a proximate generator to the embryo is the father,[40] for the father is actual [and] like that which is generated, {and that} this power [in the semen] stands in the same relation to the father as an instrument to an artisan.[41] But because such instruments are not moved except by contact with the agent that uses them, whereas in the case of this power it appears to move by itself when the proximate agent is already absent, Aristotle thought[42] it resembled, in artificial things, those extraordinary devices which appear to be moved by themselves, but in fact only acquire their motion from the agent that made them: i.e., an agent which is actually moved by itself,[43] in this case, the father.

10. Inasmuch as this power acts only by heat, and heat *qua* heat imparts nothing to what is made except a quality like itself, while it is apparent that this power imparts both the figure and the shape;[44] and because that which imparts the figure and the shape in artificial things is the form of the art,[45] it follows that this power in natural things resembles the power of the art in artificial things,[46] and the power of the soul in animate things.[47] But this power is not itself a soul, because the soul is only the entelechy of an organic body.[48]

11. When Aristotle had reached this point in his investigation of this power, he began to investigate what the nature of the power is, for it cannot be a soul, or a power like the natural powers in the elements. Since the investigation concerns the natures of the powers,[49] the only [valid] method is to investigate their functions.[50] It is apparent that the first thing this power accomplishes is the creation of the members and their formation. Thus, the physicians call it the formative power,[51] {resembling} this nutritive <faculty>[52] [in its activity], for the nutritive faculty [of the soul] {causes} what is potentially a part

of that which is nourished to become an actually besouled part of it,[53] and this power [in the semen] causes what is potentially all the parts of that which is nourished to become actually besouled, that is, actually nourished.[54] For no member of that which is nourished becomes actual unless the nutritive faculty [also] exists in it actually,[55] and the agent, as has been said elsewhere, is what bestows the entelechy of that which is acted upon: i.e., {the} form.[56] {All} this indicates that this power is what forms the members and gives them the nutritive soul, but that it itself is not a nutritive soul,[57] for the latter acts through organic instruments {while this power} has no instrument, excepting only heat.[58]

12. Because the investigation has to do with the functions of this power, there must first be an investigation of the faculties of the soul. If there is a faculty which is generated, this power is undoubtedly the generator; while, if there is some faculty which is not {generated}, it undoubtedly enters from without and does not require this power, nor does this power generate it.[59] When Aristotle had investigated[60] this matter, it appeared[61] to him that all the faculties of the soul are generated, because they require organs: that is to say, their activity cannot be accomplished except through a body, as in the case of the nutritive faculty, the {faculty of locomotion}, and the sensitive, perceptive faculty.[62]

13. As for the intellect, because it is not apparent that it possesses a corporeal organ through which it operates, as is the case with the rest of the faculties of the soul,[63] doubt has arisen as to whether it enters from without,[64] or whether {it is generated} in certain respects and in other respects enters from without.[65] Aristotle deferred the investigation of this question to the place appropriate to it,[66] and [here] he said [only] that the prevailing opinion about this is that the intellect enters from without;[67] and he decided[68] that every psychic power requires a body in which to exist.[69] And when Aristotle had speculated about the nature of this body, he said that it is a body more noble than that of any of the four elements, {and he said that it is a divine body,} surpassingly noble and excellent.[70]

14. The *Question* has been completed. Praise be to God![71]

QUESTION VII

Although Averroes' stated aim here is to prove that the arguments of *Physics* VII are not superfluous, he does far more than this. While he treats such problems as the motion of the elements and seizes the opportunity to criticize the opinions of certain predecessors, and while he offers the customary analysis of the form of Aristotle's arguments, the real interest of this treatise lies in the fact that Averroes shows so clearly his feeling for *Physics* as a structure: a structure in which everything has its place, without any superfluity, and one which is carefully designed to achieve the desired purpose

1. Still another treatise which also belongs to Averroes.[1]

2. The philosopher Averroes said: The intention in this treatise is that we demonstrate that what Aristotle proved at the beginning of the seventh book [of the *Physics*] to the effect that everything in motion has a mover,[2] and what he proved on this same point in the eighth book,[3] constitute two [distinct] inquiries, for each one of which he offered a peculiarly appropriate proof; and that neither proof is superfluous, particularly what was demonstrated at the beginning of the seventh book. Similarly, [we intend to demonstrate] that what he also proved in the seventh book, to the effect that all the things which undergo locomotion due to an external mover [form a finite series which] comes to an end at something which undergoes motion in virtue of itself,[4] and what he also proved concerning this in the eighth book,[5] neither one of these [demonstrations] is superfluous, rather {each} demonstration is appropriate in its place.[6]

3. [Averroes said:][7] We say that when Aristotle defined nature in the second book of this work as a principle in a thing by which it is moved and rests primarily and essentially,[8] he employed the terms "being moved" and "at rest" in this definition with reference to [all] four kinds of change,[9] namely, locomotion, growth and diminution, alteration, and generation and corruption.[10] In the same way, he employed the term "principle" to include that which is a soul, or that which is not a soul, or that which is an intelligence.[11]

4. This definition is self-evident there [in the second book], and peculiar to natural things, for it is self-evident that we perceive natural things to undergo change of themselves, not because of something external to them, as is the case with artificial things,[12] and, consequently, this is a clear distinction between artificial things and natural things. This definition, if [applied] to generated things, is self-evident, for the non-existent thing does not generate itself, and, similarly, it is self-evident that that which undergoes alteration does not alter itself.[13]

5. But as for that which is moved of itself in place, its {existence} is of two kinds: one kind is [a thing] moved by a principle in it called a soul,[14] and in this case its need for the mover is clear:[15] for it comes to rest when the soul no longer exists.[16] This principle exists in both plants and animals, although it is more apparent in the animal than in the plant.[17] But a certain difficulty has sometimes arisen with respect to this principle, namely, whether it moves itself in such a way that there is a mover which moves itself and which nothing else moves.[18] This difficulty arises only because Aristotle thought that everything in motion had a mover {and every mover must itself be moved},[19] and this being assumed, it follows necessarily either that the movers proceed to infinity without there being a first mover--which is absurd--, or that in the case of such moved movers, the series comes to an end at something which moves itself.[20] But if the latter is {true}, it follows that the self-mover's [being moved while] causing motion in something else is accidental.[21] For this reason Aristotle endeavored to demonstrate that everything in motion has a mover [other than itself], and that the first mover must be [absolutely] immovable.

6. Inasmuch as the same difficulty also occurs with respect to those inanimate bodies which undergo locomotion,[22] such as the four simple elements, Aristotle also investigated whether the principles by which

they are moved are identical with their substance or something in addition to their substance.[23] This was done at the beginning of the seventh book in the investigation which is peculiar to them.[24] He demonstrated that these bodies are composed of a mover which is something additional in the subject,[25] and something which is moved. For the purposes of this demonstration he laid down self-evident propositions,[26] one of them being that if there is something which is in motion of itself, it must come to rest of itself. Then he laid down another proposition, namely, that if there is something {which is set in motion because something else is in motion}, it must come to rest when [that] something else comes to rest. He concluded from these two propositions that if there is something in motion of itself, it does not come to rest because something else comes to rest. From this follows the conversion of the obverse, namely, that if there is something which comes to rest because something else comes to rest, it must be moved by [that] something else. When he had satisfied himself of the truth of this proposition, namely, that what comes to rest because something else comes to rest is moved by something else, he added to it what had been demonstrated in the sixth book, to the effect that everything which undergoes motion is a body.[27] This being assumed, and it having already been demonstrated [in the sixth book] that every body is continuously divisible,[28] Aristotle again laid down another self-evident proposition, namely, that every body is either in motion primarily (that is, so that it is not in motion due to some part in it which is in motion [primarily] of itself, or it is in motion due to some part which is in motion primarily. He considered this proposition to be true because every natural body possesses both a limited minimum quantity and a limited maximum quantity, it being impossible that there exist a body in the same genus which is smaller than the minimum quantity or larger than the maximum.[29]

7. Inasmuch as he had satisfied himself that every body is either in motion primarily (that is, so that it is not moved because of some {primary} part in it which is in motion of itself), or is in motion because it contains something in motion of this description, he considered it to be true that every body in motion *of itself* is either in motion primarily, or contains something in motion primarily. To these propositions he added [still] another self-evident proposition, namely, that in the case of anything in motion primarily, if it is assumed that a part thereof has already come to rest, the remainder must come to rest. For if the remainder continued in motion, what had been assumed to be in motion primarily would not be in motion primarily: because [the whole] would be in motion [only] at the time [the remaining part] was in motion, the whole would be in motion only due to [the motion of] a part thereof.[30]

8. Inasmuch as these propositions had been taken to be true, Aristotle deduced from this that every body is moved by something other than itself. He did this in the following manner: Every body in motion is either in motion primarily or contains something that is in motion primarily; and in the case of everything which contains something in motion primarily, or which is itself in motion primarily, when a part of what is moved primarily is assumed to come to rest, the remainder must come to rest. But everything of this description comes to rest when some-

thing else comes to rest, and everything that comes to rest when something else comes to rest must be set in motion by something else.[31] Therefore, it follows necessarily from this that every body must be moved by something else and, if this is so, that which is in motion of itself must necessarily be composed of a mover which is not a body, and something in motion which is a body.[32]

9. Having satisfied himself that that which is in motion of itself, not because of another body external to it, is of this description, he began to inquire with respect to the bodies which undergo locomotion (namely, those which move one another) whether they must come to an end at something in motion of itself, or whether it is possible for such bodies to proceed to infinity.[33] He assumed [a number of] bodies undergoing locomotion, the last of them moved by another body external to it, and that second body moved by a third, and the third moved by another, fourth body external to it; and after this he investigated whether or not it is true with respect to moved movers of this type that they proceed to infinity. He assumed that it must follow in the case of such bodies which move one another, that they cause motion and undergo motion simultaneously, that is, reciprocally, namely,[34] [so that] the mover of them must undergo motion at the same time as that [thing] in which it causes motion undergoes motion[35] (for when the mover of them comes to rest, that which undergoes motion comes to rest); and [he also assumed] that these bodies touch one another.[36]

10. Because Aristotle assumed this, it followed that an infinite magnitude, one by contiguity, must result from these bodies,[37] and [it also followed] that [the component parts of] this magnitude must be moved simultaneously.[38] But since the magnitude of the motion arises from [the size] of the magnitude which is moved with this motion, and the one magnitude which is moved with this motion, being composed of an infinite number of magnitudes, {is infinite}, it follows necessarily that the motion with which it is moved must be infinite.[39] Then, when this infinite magnitude was assumed to undergo motion in a certain time, it followed that its infinite motion must take place in a finite time;[40] but this is absurd according to what was proved in the sixth book.[41] Thus, it is necessary that bodies such as these, which both cause and suffer motion, terminate at a body which is moved by itself.[42]

11. Because it has already been conclusively proven in the preceding demonstration that everything in motion is moved by a mover, it follows that that which is in motion of itself must be composed of something in motion, which is a body, and a mover, which is not a body.[43] For if we assume the mover to be a body, it follows that that which was first assumed to be in motion of itself is not in motion of itself, but rather, [it is in motion] because of another body external to it;[44] and when I say "in motion of itself," I mean in motion through a principle [of motion] in it, not external to it.

12. When he had established to his satisfaction that [bodies in] locomotion must terminate at something moved by a principle in it which is not a body, he began, at the beginning of the eighth book, to investigate whether this primary motion, with which that which is in motion primarily is moved by the Prime Mover, can be created or not. Inasmuch as it was stated in the definition of motion that motion is the entelechy of the movable,[45] this primary motion must be the entelechy of that

primary movable body and a Prime Mover,[46] there being no other mover prior to this mover and no other movable body prior to this movable body. This being so, it follows that that which undergoes this motion must be eternal, for if it were generated there would have to be motion before the primary motion, generation being a motion or the consequence of a motion.[47] But in this case an absurdity must follow, namely, that the motion which was assumed to be primary is not primary.

13. When Aristotle had verified that that which is moved with the primary motion is eternal, he went on to investigate whether it is possible for something like this thing in motion which is moved with the primary motion to be in motion at one time and at rest at another. Inasmuch as everything set in motion after being at rest has a motion before it which necessarily causes it to be set in motion at the time it begins to move after being at rest, for that reason it is impossible that any created motion should exist without another motion before it: because every motion is created only by a mover and something in motion.[48] This being so, when the {eternal} thing in motion is assumed to be in motion after having been at rest, it follows that it cannot be a first thing in motion, nor can its motion be a first motion, nor its mover a prime mover. All this, however, is in contradiction to what was assumed.[49] If a created motion lacking a motion prior to it in time[50] is assumed, a motion would exist without a mover which is created with respect to causing motion; while if one assumes that there is a mover which is created with respect to causing motion, it follows that there must be a motion before the posited motion.[51] Thus, the eternal[52] thing in motion must necessarily be moved with the primary motion {and by the prime mover} which is, undoubtedly, eternal with respect to causing motion.

14. What the Mutakallimûn of our coreligionists[53] have assumed, that it is possible for the mover that acts voluntarily to cause a first motion which has no motion before it either in the mover itself or in that which is set in motion by the mover, is an absurd assumption. For it will be demonstrated from the accidents[54] of that which is set in motion of itself after being at rest that it is not set in motion after being at rest unless another motion has already preceded [this setting in motion], either in its body or in its soul.[55]

15. You ought to know that the method which we have followed in establishing the eternity of motion is the method of Aristotle himself at the beginning of the eighth book, not the method which John [Philoponus][56] understood to be Aristotle's, [Philoponus' understanding] being identical with the method al-Farabi, in his book *On the Mutable Existences*,[57] understood to be Aristotle's. For this reason, the investigation [undertaken] in that book, with respect to establishing [the existence of] an eternal motion, was confused; for [what] al-Farabi understood from [what] Aristotle [wrote] at the beginning of the eighth book was that Aristotle only employed this proof: that the potentiality for a given motion must precede that motion in time. [This is] because, if that thing which possessed the potentiality [for motion] is assumed to be set in motion with a created motion, a potentiality for that [created] motion must precede the motion, so that it is altogether impossible for us to assume an initial motion if there is no motion before it.[58] But this conclusion [that there is no initial motion] fol-

lows necessarily [only] from the assumption of an eternal motion, for it is not[59] something which follows from the nature of motion itself; rather, [the lack of an initial motion] is accidental to motion because there is an eternal motion.[60] If we were to assume that this [lack of an initial motion] {belongs to motion} essentially, it would follow that there exist movers which cause motion in one another without having an initial mover, as Aristotle maintained in the third book [of *Physics*]:[61] that if there were a motion prior to the created motion, that is, such that its creation were a consequence of another motion essentially, and that other motion a consequence of still another, so that the series proceeded to infinity, it would follow that the last motion could not exist except after the completion of infinite motions; but if the first does not exist, the last cannot exist at all.[62]

16. This is what Plato and his successors, of those who professed his opinion, thought. They considered it impossible that every motion should have motion before it, and what they thought about this is true if this [relation between the motions] is assumed to be essential, but false if it is assumed to be accidental, as is clear from the positing of an eternal motion in accordance with what was demonstrated at the beginning of this [eighth] book by the method we have mentioned. Because what is accidental there was taken to be essential, doubts and perplexities which are hard to settle have arisen among Aristotle's opponents with regard to this question: for one who accepts the existence of an eternal mover must accept the fact that there is an eternal motion; and one who assumes that there is an eternal motion must assume that there are {motions which have} no beginning--but if this is assumed to be accidental, the difficulties are resolved.[63]

17. {Thus,} al-Farabi, in his book *On the Mutable Existences*, had to investigate all the species [of motion] in which motion after motion [without an initial motion] exists, so that the species in which this is possible might be distinguished from those in which it is impossibly absurd.[64] His discourse was confused because he thought that Aristotle defined motion at the beginning of the eighth book only so that he might demonstrate that the potentiality for motion is prior to motion in time; but this is evident with respect to every created thing, whether that created thing is a motion or something in motion.[65]

18. So, too, {what} John Philoponus thought: that this method was Aristotle's method of establishing that there is an eternal motion in contradistinction[66] to the motion in virtue of themselves[67] of the four elements; for it appears that the potentiality for motion does not precede the motions of these elements in place, except when the elements are compelled to be motionless.[68] Because al-Farabi conceded that this method which John Philoponus thought was Aristotle's {was <indeed Aristotle's> method} of proving that there is an eternal motion which is in contradistinction [to the motion of the elements], {it appeared [to him] that the potentiality} which precedes this motion [of the elements] is the potentiality which exists in the body from which these [elemental] bodies are generated.[69] {But} [actually] Aristotle presented the definition of motion [at the beginning of the eighth book] only so that it might be demonstrated from it that every motion is in that which is moved,[70] and that motion never ceases to exist because something in motion never ceases to exist.[71] This follows properly from the defini-

tion of motion.[72] But what these [aforementioned] men thought does not follow properly from the definition of motion, for the potentiality precedes both the motion and that which is in motion.[73]

19. Having explained this, let us return to the order which Aristotle followed in this book. We say that when he had arrived, by a proof we have mentioned, at the existence of an eternal motion and an eternal mover, because of the cogency of an assumption of a primary motion for things which undergo locomotion, he then demonstrated [the existence of] this [mover and this motion] on the basis [of a proof from the nature] of time as well,[74] his demonstration on this point being clear.[75] Then he brought up certain difficulties[76] occurring with respect to his assumption[77] [that every created motion is preceded by motion, one of them] to the effect that it is possible to assume something which is moved with a created motion which has no motion preceding it, namely, what appears to be the case with the animals. For it is thought that the animals undergo motion after perfect rest, and that no other motion precedes their locomotion.[78]

20. Because he had reached this point [in his argument], Aristotle thought that the best way to demonstrate this thesis[79] was to investigate the qualities of all existent things with respect to motion and rest.[80] Consequently, he demonstrated, by showing that we perceive some things to be in motion at one time and at rest at another, that it is possible neither for all of them to be in motion, nor for all of them to be at rest, nor for them to be divided into only two classes in this respect,[81] so that some of them are in motion perpetually and others at rest perpetually, The question remained, then, whether all existing things are of this description, that is, in motion at one time and at rest at another, or whether there are some in motion at one time and at rest at another, and there are also some in motion perpetually and [others] at rest perpetually.

21. Because he intended to investigate this [latter] genus,[82] he began to investigate[83] how the case stands with respect to the motion of the heavy and light [bodies].[84] [This was necessary] because it had already been demonstrated at the beginning of the seventh book that nothing exists which is in motion of itself, that is, in the sense that its mover is identical with that which is moved. But there is also a question with respect to the motion of the elements: whether they are more properly to be ascribed to the class of beings moved by external movers, or whether they may more properly be ascribed to the class of beings in motion of themselves through principles in them, as is the case with animals.[85]

22. The difficulty concerning these two [kinds of beings][86] must be discussed so that, from this discussion, it may become clear how the case stands with respect to that which is in motion primarily of itself: whether it is of the genus of beings which are moved naturally, that is, if it is a being in the rank of the elements; or whether it is of the genus of beings which are moved by themselves, [the genus] of beings endowed with souls. He began to investigate this by investigating the elements, and he demonstrated that they are more properly to be ascribed to the class of beings which are moved by things external to them than to the class of beings which are moved by themselves.[87] The preceding question with respect to these two [kinds of beings] was resolved for

him, and it was shown to be true that they are moved by that which is other than themselves, and that it is not possible for that which is in motion primarily to be of the nature of [either of] these two [kinds of beings].

23. Because some of the commentators did not understand this purpose, they thought that what was demonstrated here [in the eighth book] about the elements, namely, that they are moved by a mover other than themselves, was more persuasive than what had been demonstrated [about them] in the seventh book.[88] But there Aristotle speculated about them only in so far as they are moved by themselves, and on this basis he constructed the preceding proof concerning the eternal motion and the eternal mover.[89] Here [in the eighth book], however, he speculated about the two of them[90] only in so far as they are moved essentially, that is, in so far as they are moved by a mover external to them; and in so far as their being moved by themselves is accidental to them,[91] while that which is essential to them is that they are moved by something external to them; and that, because of this, that which is in motion primarily cannot be assumed to be moved by some natural principle in it which resembles the principles in these [elemental bodies]. For {it is evident}, as Aristotle demonstrated, that they are not in motion of themselves when they are generated in their natural places,[92] and that when they are generated in places outside their natural places, every part of them that acquires a part of the form [of a particular element] also acquires a part of the place [proper to that element];[93] so that, when their generation has been completed, they attain [their ultimate perfection] in their natural places,[94] unless there is something there that prevents them from doing so.[95] But if the elements are assumed to be moved to their [proper] places [after having been prevented from doing so], their motion of themselves in this case is accidental, for that which <removed what> withheld them from motion is something accidental to them.[96]

24. When he had satisfied himself by proof that this, that is, the elements, is in motion of itself only accidentally, it appeared to him to be true that it is moved essentially by an external mover. Furthermore, if it were in motion of itself, it ought to come to rest of itself, as is the case with those beings which are in motion of themselves, namely, those beings which possess souls.[97]

25. He had already satisfied himself that the [series of] bodies which cause motion and suffer motion reciprocally can come to an end at something of this kind which is in motion of itself,[98] according to the proof written at the beginning of the seventh book:[99] namely, that if they do not come to an end at something in motion of itself, an infinite magnitude exists which is moved with an infinite motion in a finite time. For example, the waves move the ship, and the wind the waves, and the wind is in motion of itself accidentally, but is in motion essentially due to an external mover. Thus, the bodies such as these, which both cause and suffer motion, must terminate at things in motion belonging to the genus of the elements.

26. [But] it is also possible for us to assume that the generation of the last [in a series] is first, and when that last one is generated it is moved by itself and the mover which was before it, that is, the first, is corrupted.[100] We can also assume {a third} whose relation to

the second is like the relation of the second to that which was generated first, and this can proceed infinitely, as in the case of the generation of man from man. Thus, it might be thought that such things in motion[101] do not terminate at a first mover.

27. It therefore followed necessarily that the proof that Aristotle wrote in the seventh book could not suffice [with respect to establishing] that those things which are in motion in place terminate at something which is in motion of itself. In order to prove this, he followed a method of proof different from the method he followed in the seventh book: namely, {when he assumed}[102] [in the eighth book] that things such as these, which cause motion and suffer motion reciprocally, do not exist simultaneously with respect to causing motion in one another; {and when these things were assumed} to be infinite [in succession], or were assumed to be so essentially (that is, when this infinity is not assumed to be accidental to them because there is an eternal mover which moves itself).[103] But when these things are assumed to be infinite in succession essentially, there is no first mover in the series; and if there is no first term in the series, there is no last term. For what has no beginning has no end, and if a last term did exist in things such as these, it could not exist until an infinite series had come to an end, which is absurd.[104]

28. Because he had demonstrated that all things in motion must necessarily terminate at something in motion of itself; {and because} [it had been demonstrated] that the reason for things in motion of themselves[105] being at rest at one time and being moved at another is something which is a consequence of motions generated in them, either in their bodies or in their souls; and because it had already been demonstrated that the Prime Mover of the universe is immovable and, similarly, that the first thing in motion is not subject to [any other] change,[106] inasmuch as the motion which results from the two of them is eternal: it follows necessarily that the first thing in motion of itself, which is composed of a mover and something in motion, is not subject to change either because of [a motion produced in] its magnitude or because of [a motion produced in] its soul.[107]

29. Inasmuch as the cause of the change of the magnitudes[108] of the beings in motion of themselves in the sublunary world is either their bodies, because they are compounded of matter and form, or the motive powers in them, with respect to which motion exists only accidentally (because although they are not bodies, they inhere in bodies without which they cannot exist), it follows that that first thing in motion must be simple; for if it were compound, it would be susceptible of change.[109] {Thus, by the conversion of the obverse it follows that what is not susceptible of change is simple. This is why Hippocrates said that if man were one thing he would not suffer pain.[110] Because it has already been demonstrated that the Prime Mover is not susceptible of change}, either essentially or accidentally, and [because] it is also evident that it is not a body, it must, further, be true of it that it does not inhere in that which is moved by it. For if it did inhere in that which is moved primarily by it, as is the case with the souls of animals in the sublunar world, that which is moved primarily by it would not be simple, nor would that mover be a prime mover. It follows, therefore, without question, that both the Prime Mover and the first

thing in motion are simple.[111]

30. When we have assumed that there is a mover and something in motion of this description, the preceding proofs do not suffice to demonstrate that everything in motion is moved only by a mover other than itself, for those proofs are valid only with reference to things in motion {which are compounds, and things in motion whose existence is in a subject}. This is clear from what was written at the beginning of the seventh book, for the question there concerns only the elements. It is also clear from what was written in the eighth book before reaching this point, to the effect that the elements are set in motion only by an external mover, and that the case with respect to those things which possess souls is clear.[112] Furthermore, he assumed from this [latter inductive] investigation that everything in motion has a mover, without examining in that [investigation] the simple moved thing. For all these reasons, Aristotle found it necessary to return to the beginning[113] and demonstrate that there exists no body, whether simple or compound, in which that [part] of it which is in motion is identical with the mover. Thus, he maintained that that which is in motion primarily must necessarily {either be moved by [what is at] rest,} or be in motion of itself, that is, so that the mover in it is identical with that which is in motion. He said, "For if one assumes that there is something in motion primarily, it is more properly moved by itself than by something else, for that which is in motion of itself, if it exists, is more properly [to be described] by the term 'primary'."[114] Then he began[115] to investigate whether it is possible for any body to exist, whether simple or compound, in which the mover is identical with that which is in motion. [If it is assumed that this is possible,] all the impossibilities which he mentioned in this book ensue from this assumption, and from all the impossibilities it follows that the mover cannot be identical with that which is in motion.[116]

31. Inasmuch as what he assumed here, to the effect that the first thing in motion is a simple body, had not been demonstrated sufficiently in the preceding proofs (for the proof stated at the beginning of the seventh book is demonstrative only with respect to bodies compounded of matter and form, and the earlier proof [in the eighth book] was constructed only on the basis of the investigation derived from the sensible bodies in the sublunar world: those things in motion which possess souls, and the elements); and because it had not yet been demonstrated that there is a simple body which is not compound: for this reason, Aristotle began here to investigate this question from the beginning.[117] He explained by a general discourse[118] that it is not possible for anything to move itself, whether it is simple or compound, for in that case the mover would cause motion in so far as it was in motion, and it would exist in so far as it was non-existent. When he had completed the verification of this proposition and had negated its contradictory {in all respects}, he enumerated the ways in which it is possible to conceive that something moves itself.[119] He began by saying that that which is in motion primarily must inevitably be one of two things: either it must be something set in motion by what is [itself] unmoved, or it must be something which moves itself. He said, "For if there is something which moves itself, it would more properly be [described as] in motion primarily than that which is moved by something

else."[120] When he had assumed this, he [then] assumed, according to what had been demonstrated in the sixth book,[121] that that which is in motion is a body, whether that body is simple or compound. He further assumed that if this body moves itself, either the whole of it moves itself (and this is absurd because then the mover would be identical with that which is in motion), or one part of that body moves another.

32. These parts which move one another must inevitably move one another either with a rectilinear motion or with a circular motion.[122] If the motion is rectilinear, then it must be either finite or infinite. If it is finite, the first absurdity must follow, namely, all the impossibilities which he mentioned in this book. If it is infinite, it follows that there exists an actually infinite number of magnitudes in the finite magnitude,[123] and it will not be moved primarily.[124] Furthermore, when the first does not exist, the last cannot exist.[125] If the parts are moved with a circular motion, the first absurdity returns: namely, that that which is moved by something else moves itself, and that the prior is posterior.[126]

33. When [it had been shown that] it is impossible for this body in motion of itself which is simple, not compound, to move itself in its totality itself, or to move itself because of its parts (because from the latter follows the same impossibility that follows from the impossible assumption that it moves itself in its totality because of [the totality] itself, and not because of its parts), then it had already been [shown to be] untrue that any body at all exists that moves itself, whether we assume that body to be simple or compound. Thus, he had completed this general proof:[127] namely, that every motion terminates at something in motion of itself, and that everything in motion of itself is composed of an unmoved mover which is not a body, and something in motion which is moved by that mover.

34. Inasmuch as it is self-evident that there are many things in motion of themselves which are in motion at one time and at rest at another, and which exist at one time and are non-existent at another, and some of which are causes for others: it is evident with respect to these things in motion of themselves that the movers in them, because they are not bodies, are in motion accidentally; for these bodies exist at one time and are non-existent at another, and are at rest at one time and in motion at another. Inasmuch as these things which are in motion of themselves are [in motion] accidentally because other movers are prior to them,[128] it follows that each one of these must terminate at what is moved by itself essentially, whether there is one such thing or more than one.[129]

35. Inasmuch as it is evident that these things in motion [of themselves accidentally] are eternal in species, and that this [eternity] is accidental to them because they have no first (that is, of their species) which is the principle of their motion, it follows that these eternal things must only acquire [eternity in species] due to something in motion of itself which is not subject to change accidentally in any of the four species of change. This must undoubtedly be that which is in motion of itself essentially, that which no motion at all, or mover, or thing in motion precedes.[130] This may be one or many, and if many, they may be finite or infinite [in number]. But it is apparent that its being one is all that is necessary for the existence of these things in

motion which are eternal [in species], and if more than one exists, this is because it is better, not because it is necessary. Moreover, if more than one exists, it is better that they should be finite [in number], for the same result follows from the [assumption of a] finite [number] as follows from the [assumption of an] infinite [number], and nature does not act in vain.[131]

36. Inasmuch as he had reached this [point in his discussion] concerning those things which are in motion of themselves, and [inasmuch as] his investigation concerned all species of things in motion and at rest, it became clear to him {because} of this proof that some things are in motion at one time and at rest at another, while others are in motion perpetually, and [still] others are at rest perpetually (namely, those around which the things perpetually in motion are moved; for nothing can be in motion without something at rest around which it is moved).[132] Also [it became clear to him] that the things which are in motion at one time and at rest at another are of this description only because of that first thing in motion [which is moved] by itself essentially. For it is not necessary that each one of the things[133] [in motion] should be moved perpetually, inasmuch as that [first thing in motion] is close to them at one time and [so] causes them to move, and is remote from them at another time and [so] does not cause them to move.[134] Thus, it is said that that which is in motion of itself does not necessarily cause motion,[135] whether it is in motion {of itself} essentially {or accidentally}.[136]

37. Thus, from the existence of things in motion in the sublunar world which are eternal [in species], he had verified the existence of something in motion of itself which is eternal, the mover in which is subject to motion neither essentially nor accidentally. [And he established] that that which is moved by this mover is undoubtedly eternal (while the compound thing is not eternal), and that the mover in it does not inhere in that whose motion is caused by it: for were this the case, the mover would be subject to motion accidentally, and [that in which it causes motion] would not be the first of things in motion of themselves.[137] In this way, therefore, he established to his own satisfaction the existence of an eternal motion on the basis of the existence of that which is in motion of itself.

38. The first proof, namely, what was demonstrated at the beginning of the eighth book, is a *demonstratio per signum*; for there it was demonstrated that there is an eternal mover because there is an eternal thing moved, and that the thing moved is eternal because there is an eternal motion.[138] It was also demonstrated there that they are eternal because they are primary. The proof used here, however, is a *demonstratio per causam*, for this proof demonstrates the existence of an eternal motion on the ground of the existence of something moved by itself which is eternal.[139]

39. The first proof is analogous to the proof one uses to demonstrate that the shape of the moon is spherical because the increase of its light is in the lunar shape, while the second proof is analogous to the proof one uses to demonstrate that because the moon's shape is spherical, the increase of its light is in the lunar shape.[140] Undoubtedly, this second type of proof is better than the first, and this is the reason that Aristotle introduced {this} second proof regarding the eternity

of motion after [he had already given] the first proof; and he did this also so that one might understand from this second proof the causes of the [three] species of beings as regards motion and rest.[141] This was his primary intention, and whatever else was demonstrated in this proof is by way of pure profit.

40. When this excellent thesis had been demonstrated to his satisfaction by this excellent proof, he began to ask which motion is this [eternal] motion.[142] Because it is evident that that which is in motion of itself exists only [in things] in locomotion, that is, the motion which is called "translation," it was demonstrated by him that this [eternal] motion is a translatory motion. Because the motion of translation is either rectilinear or circular, and it had already been demonstrated that rectilinear motion is finite because it is from an opposite to an opposite,[143] it became clear to him that this motion must be circular, and one and continuous (not contiguous[144] to another motion, nor successive to another motion),[145] and that it is not a motion which turns back over a straight line.[146] When all this had been demonstrated to his satisfaction, it became clear to him that this motion, which is a translatory motion, is continuous, belonging to one thing in motion and one mover.[147]

41. When he had demonstrated this, he undertook to demonstrate also, by induction,[148] that it is at this motion that the rest {of the species} of the motions terminate, and that this motion is prior to them in nature, and time, and [perfection of] existence[149] [in] the world as a whole, [not] in the individual.[150]

42. Inasmuch as it had been demonstrated that this Prime Mover is not a force in matter because it is not subject to motion [even] accidentally,[151] Aristotle thought it best to demonstrate the existence of this characteristic in the Prime Mover on the basis of the activity peculiar {to it}, namely, <that that which is moved by this mover> is moved[152] with a perpetual motion.[153] He assumed that every perpetual motion is infinite, so that it results only from an infinite force.[154] Then he assumed that this infinite force must inevitably be either in a body, or not in a body (that is, inhering in a body, or not inhering in a body). He next assumed a self-evident principle: that every force in a body, whether that body be simple or compounded of simple bodies, must be divided when the body [in which it inheres] is divided.[155] When he had assumed these two [latter] propositions, he deduced from them that no infinite force can exist in a body {because such a force would have to exist in an infinite body, and the existence of an infinite body is absurd}.[156] It follows, therefore, that this force does not exist in a body.[157]

43. With respect to this point, he then employed many proofs by way of revealing [the truth] through a *demonstratio per impossibile*.[158] Thus, he assumed an infinite force in a finite body, and from this followed obvious impossibilities, as he wrote at the end of this book:[159] one of them, that in this body the force of the whole would be [equal to] the force of the part;[160] another, that this body would be moved in an instant;[161] another, that there would be an infinite motion greater than another infinite motion, for if the motion of the <part> is smaller than the motion of the <whole>,[162] and the whole and the part are both moved with an infinite motion, {it would follow that one infinite

motion} is greater than another infinite motion.[163] This is the end of what Aristotle intended to demonstrate in this [eighth] book.[164]

44. It is obvious also from the fact that the first movable body is the recipient of eternal motion, that it is simple, and that it is not compounded of form and matter;[165] {for that} which is compounded of form and matter is subject to generation and corruption, while it is impossible that this [eternal] motion, whose existence is primary and essential, should be successive to another motion or contiguous to another motion.[166] Rather, that which is moved with this motion must be one and simple.[167] Aristotle does not [attempt to] prove this point with regard to the first moved body [here] because it is exceedingly obvious[168] from what had [already] been demonstrated concerning the [Prime] Mover;[169] and because it had already been demonstrated that if [this first] moved body were compounded of matter and form, it would not be in motion of itself except accidentally, nor would it be moved primarily by the Prime Mover; rather, it would be moved through the mediation of another mover.[170]

45. Certain men have thought, because it was demonstrated that this [first] moved body, namely, the celestial body, is eternal, and because it was also demonstrated that it possesses a soul,[171] that this body does not require a separate [i.e., incorporeal], external mover.[172] They {ignored} the fact that assuming it to be compounded of a mover which is not a body, and a body, was an assumption which contradicted the assumption that it is eternal: for the bodies of this description[173] are compounded of matter and form, and every compound of matter and form exists potentially at one time and actually at another, so that it is necessarily generated and corrupted. But if two things are assumed which have no {difference} at all between them, either potentially or actually,[174] this is a self-contradictory assumption: namely, that what was assumed to be one is not one.[175] Furthermore, this simple body does not require the Prime Mover in order to be eternal; rather, it requires the Prime Mover only in order that it be moved with an eternal motion by an eternal mover: for it cannot possibly move itself, according to what has been demonstrated, that is, in the sense that the mover is identical with what is moved.[176]

46. It is evident from this that every body, whether simple or compound, is finite with respect to intensity in its ability to cause motion,[177] for if this were not so, the celestial bodies would destroy all the moved beings in the sublunar world.[178] Thus, it has been established that every body has a motion finite with respect to intensity and force, and that this [intensity] is a consequence of the size of its magnitude,[179] and that the motion [with which it is moved] by its mover is finite with respect to velocity.[180] If this were not so, the relation of any incorporeal mover whatsoever {to} any moved thing whatsoever could be any relation at all; or there might be no relation there,[181] so that it would follow that there could be motion in an instant.[182] For this reason, everything in motion [must be moved] by its mover with a motion finite in velocity, and in a finite time;[183] and the mover cannot move any chance thing moved, nor can it move it with any chance measure of motion.[184] If this were not so, a mover might cause the same motion in the larger [object] as in the smaller,[185] and this is absurd.

47. This, then, is the sum and substance of what Aristotle demon-

strated in this [eighth] book, and this is the method which he followed. Many men did not understand this method and followed another method, full of confusion, as did al-Farabi in *On the Mutable Existences*, so that it was hard for them to resolve the objections which were raised against them about the eternity of motion--objections which the Mutakallimûn of our religion took over from Plato and those speculative thinkers who followed his doctrine, such as John Philoponus and others. But for one who understands these matters on the basis of this book, it is easy to resolve the objections which they introduced concerning the eternity of motion, and particularly the objection raised by John Philoponus which is the most important and the strongest of the objections which may be raised in this regard.[186] [This objection may be stated in the following way:] He said that Aristotle and all his colleagues thought that every body has a finite force; how, then, can the celestial body, which is finite with respect to force, acquire an infinite force from the Prime Mover, unless it is true that that which ought of its own nature to be corrupted can acquire eternity from an eternal being?[187] Aristotle, however, had already demonstrated that the eternal can have no potentiality for corruption, and he did this at the end of the first book of *De Caelo et Mundo*.[188]

48. The answer to this objection is that bodies are of two kinds, simple and compound, as was explained [earlier] in this book.[189] The simple body is of infinite force with respect to existence and duration, for the cause of finitude in this respect is composition; but it is of finite force with respect to the velocity [of motion] and the intensity of [its ability] to cause motion.[190] It is finite with respect to velocity because no motion can exist except in time, and this has already been proven demonstratively.[191] It is finite with respect to intensity {because} every body is finite with respect to magnitude, whether that body is simple or compound. But the bodies which are compounded of matter and form are finite in three {respects}: namely, in their duration, their velocity, and their activity with respect to intensity.[192]

49. This is what we desired to establish in this treatise. Praise be to God alone, blessed and exalted be He!

50. The treatise has been completed. Praise be to God!

QUESTION VIII

In this treatise, Averroes defends Aristotle against the charge of using invalid premises in his demonstrations. He begins by analyzing a specific argument, moves on to the charges raised against it, and then broadens his defense of the premises of this argument to cover Aristotle's general practice. Averroes' defense of the practice follows a line already laid down by Aristotle himself in *Physics* VII, and turns on the distinction between possible falsehoods and impossible falsehoods.

1. Still another treatise of Averroes.[1]

2. Averroes said that the Philosopher has demonstrated that there must necessarily be one eternal, continuous, circular motion prior to the rest of the motions, and that that [sphere] which is moved therewith cannot undergo change in its substance,[2] and that the mover which causes the motion of this movable sphere is not subject to change at all, nor can it undergo locomotion either essentially or accidentally.[3] Inasmuch as sense perception does <not[4]> discover that there is any motion to which the description arrived at by the [aforementioned] demonstration can apply except the diurnal motion, it must be true that the diurnal motion is the motion described, and it may be demonstrated further that the sphere of the diurnal motion is the first of the beings which undergo motion, and that the action of this sphere, which is its motion, is an action which never ceases. In addition to this, Aristotle also wanted to demonstrate that the mover which causes that moved [sphere] to be moved with such a motion is neither a body nor a force inherent in a body.[5]

3. In order to demonstrate this, Aristotle made use of two methods of proof, one of them being a demonstration that it is impossible for an infinite[6] action to proceed from a finite force, [while the second is a demonstration][7] from which it follows that this infinite action must necessarily belong to a force which does not inhere in a body at all.[8] Taking the conclusions of these two arguments as premises, he deduced from them that the mover which causes this [infinite motion] is not a force in a body but, rather, a force separate from every body.[9] It is through this separate [incorporeal] force that this greatest body,[10] which is in turn a condition for the existence of every other body,[11] has its existence.[12]

4. In order to prove these two fundamental principles, Aristotle made use of certain propositions. For the proof of the first principle, he made use of two propositions, the first stating that the greater of two forces causes motion for a longer time,[13] and the second stating that whatever [finite] force[14] we originally assume, we can then assume a greater force.[15] When he had laid down these two propositions, he was able to conclude that a finite force capable of moving a certain movable object for an infinite time cannot exist. For if we were to assume a greater finite force,[16] it would follow that this greater force must cause motion for a longer time than than the lesser force, inasmuch as the greater force causes motion for a longer time. There can be, however, nothing greater than what is infinite.[17] Therefore, this line of argument results in an impossible self-contradiction. This impossible conclusion follows because we assumed two propositions, about the first of which, namely, that there is a finite force which causes motion for an infinite time, there is some doubt; while the second proposition, namely, that whatever finite force we take, we can assume a greater finite force, is known to be possible. The impossible conclusion must follow from one of these two propositions, but it cannot follow from the proposition known to be possible; as demonstrated in *Prior Analytics*, an impossible falsehood cannot follow from a possible falsehood.[18] Undoubtedly, then, the only remaining alternative is that the impossible falsehood follows from the proposition about which there is ground for doubt,[19] that is to say, the proposition which states that there is a

finite force which causes motion for an infinite time. But that from which an impossible conclusion follows is itself impossible. This assumed proposition, therefore, must be impossibly false, while its contradictory,[20] which states that every infinite action proceeds only from an infinite force, must necessarily be correct.

5. After Aristotle had established to his satisfaction that the motive force which causes this infinite motion is an infinite force, he began to demonstrate as well the second of the theses[21] [under consideration] which states that a force of this description cannot inhere in a body, by again making use of a self-evident proposition. This proposition states that if we imagine a certain movable object to be moved by two forces, one greater than the other, and [it is further assumed] that the greater force causes a faster motion, and that the proportion between [the velocities of][22] the motions is the same as the proportion between the forces,[23] then, if we [also] assume one of the two forces to be finite while the other is infinite,[24] the movable object must be moved in an instant by the infinite force, and this is impossible.[25] Moreover, should one [be tempted to] think that the movable object is moved by the infinite force in time, an impossible conclusion will follow in this case as well, namely, that one body could be moved by both a finite force and an infinite force, over one and the same distance, in the same time. To demonstrate this point, let us assume that the finite moved body is body B, and let us assume that it is moved by the infinite force signified by the letter A.[26] Furthermore, let us assume that a finite force, signified by the letter C, moves body B over an identical distance in a longer time. Moreover, let the finite force, signified by C, be added to [continually].[27] From all this it follows that the finite force C must necessarily attain such a magnitude as to move body B over a certain distance in the same time as that in which the infinite force A moves it.[28] But this is an impossible self-contradiction. This impossible conclusion must follow either from the proposition in the argument about which we are in doubt, or from the proposition known to be possible which was assumed in this argument. But, as we said before, the impossible cannot follow from the possible. The only remaining alternative, therefore, is that the impossible conclusion follows from the doubtful proposition in the argument.[29]

6. This kind of *demonstratio per impossibile*,[30] namely, the direct syllogism[31] which infers a self-contradiction from possible propositions in the syllogism and from propositions about which there is ground for doubt, was frequently used by Aristotle in the physical sciences, and, indeed, he employed it in many places.[32] A certain group of men disputed with Aristotle about the force of this kind of demonstration. They maintained that it is a corrupt demonstration because the possible propositions that Aristotle used are impossibly false with respect to that which he sought to demonstrate by means of those propositions.[33] They said: "This is the case, for example, with regard to the force of the largest sphere: if[34] we assume it to be finite, then there can exist no greater force unless there exists a larger body[35]--which is impossible. Thus, if we assert that whatever finite force we assume, we can assume a greater finite force,[36] this is a proposition which is partially false and wholly impossible.[37] And were we to say that for every finite body it is possible to assume a finite force that moves it,[38] a

similar situation would exist: for when we assume that the motion of the sphere is infinite, it follows that the force which moves the sphere must be infinite.[39] Thus, our assumption that there can be a finite force which moves the sphere is both false and impossible."

7. Among those who hold this opinion[40] are many of the modern philosophers we have encountered who follow the doctrine of Avicenna. They think that this is the opinion of Avicenna, and that it is the opinion to which he inclined in the *Oriental Philosophy*, and that there is no being which is not a body, subsisting in itself, separate from the celestial bodies, which is itself a principle of those bodies and what exists through them, as was the opinion of the Peripatetics.[41] With all this, they think that this was the opinion of the Philosopher himself,[42] and that what is to be found in his books is only by way of that concealment which was the custom of the ancient philosophers,[43] who followed it because of the prevalence of ignorance.[44] But their opinion is the worst doctrine in philosophy to which one can adhere, and the most harmful of the newfangled[45] doctrines which flourish in philosophy.

8. Let us, then, examine possible propositions such as the aforementioned. We maintain it to be evident from what was said in *Prior Analytics* that an impossible conclusion cannot follow from a possible premise. For when two things are connected necessarily, that is, so connected that if the first exists the second must exist, it is obvious that if the first is possible the second cannot be impossible: for [were the second impossible,] the first would be impossible and not possible, because the second was assumed to follow necessarily from the first.[46] Therefore, if the second is impossible, the first must [also] be impossible; but we have already assumed the first to be possible--an impossible self-contradiction. It is evident that in the case of the syllogism in which both premises are possible, no impossible conclusion whatever can be inferred. If there is, however, only one possible premise in the syllogism, an impossible conclusion cannot, indeed, be inferred from that possible premise. Rather, if such a conclusion is inferred, it must follow from the impossible premise in the syllogism. It makes no difference whether the possible premise is possible essentially, but impossible accidentally (e.g., in a certain place), or whether it is possible and in no sense impossible. By this I mean to say that an absolutely impossible conclusion cannot be inferred from the premise which is essentially possible but accidentally impossible. Rather, if the first of two necessarily connected things is of this description, it follows that the second must be of that very same description.[47]

9. This having been settled, it appears that the impossibility that follows from the syllogism which Aristotle used in this demonstration must be either an absolute impossibility, in which case it cannot be inferred from an accidentally impossible but essentially possible proposition; or an accidental impossibility, in which case it can be inferred from an accidentally impossible proposition. [We must add this latter stipulation] because the propositions which Aristotle postulated in these demonstrations are not possible absolutely,[48] for if they were, there would be no ground for doubt, and no one could become confused [about the validity of these demonstrations].

10. It is obvious that the proposition that Aristotle assumed in the

first proofs,[49] namely, his assumption of a body larger than the heavens, is possible with reference to a body *qua* body, but accidentally impossible with reference to the heavenly body by reason of the fact, for instance, that all other bodies are contained by the heavenly body, as Aristotle said when he gave the cause of the world's being one.[50] It is [also] obvious that the impossibility which follows from the syllogism in which [this] premise was assumed is an absolute impossibility, namely, the impossibility of the greater force not causing motion for a longer time. Thus it is clear that this impossible conclusion does not follow from the syllogism from which it was inferred because an essentially possible, but accidentally impossible, premise was assumed in the syllogism: namely, the proposition stating that whatever body *qua* body [is postulated], it is possible to assume a larger body. Rather, it must follow from a premise which is impossible by necessity, seeing that the conclusion is [absolutely] impossible. Inasmuch as this conclusion certainly does not follow from the aforementioned premise, it must follow from that premise in the syllogism about which there is ground for doubt: namely, that in bodies there can be a finite force which causes motion for an infinite time. This premise, therefore, is of necessity absolutely impossible; since what follows from it is absolutely impossible, the proposition itself must be absolutely impossible.

11. What is to be said with reference to the second proof,[51] and with reference to many of the places in which Aristotle made use of this kind of demonstration, is of the same nature. It was by demonstrations of this same type that Aristotle showed, because of the speed and slowness which exist in motion, that time and, in the same way, magnitude are divisible into what is always divisible.[52] For, inasmuch as he assumed that for every motion there exists both a faster and a slower motion; and that the faster motion always takes place over an identical distance in a shorter time; and that the slower motion takes place in the same time over a lesser distance: it followed that the faster motion always divides the time and that the slower motion divides the distance.[53] Thus, if a motion can exist in a time without divisible magnitude,[54] [and] it is possible to assume that a faster motion exists, an impossible conclusion must follow: namely, that what is not divisible is divisible.[55] But that from which an impossible conclusion follows must itself be impossible: in this case, our assumption of an indivisible time. It is obvious that the premise in this syllogism which produces the self-contradiction, [when taken] together with the proposition whose negation is intended, is essentially possible but accidentally impossible, because no motion faster than the diurnal motion [actually] exists. But inasmuch as the conclusion inferred from this syllogism is absolutely impossible, we know that this conclusion cannot be inferred from the proposition which is essentially possible but accidentally impossible. The impossible conclusion, therefore, can be be inferred only from the [other] assumption posited, namely, the assumption of an indivisible time.

12. Again, by this [same] kind of proof, Aristotle demonstrated that the mover in anything moved primarily is other than that which is moved. He did this by assuming that, with respect to anything in motion, it can be supposed that a part of it comes to rest, and that, when the part comes to rest in such a moved thing, the whole comes to rest.[56]

13. And God directs to certainty. The *Question* has been completed. Praised be God!

QUESTION IX

Here Averroes argues against Avicenna's belief that there are beings possible in themselves, but made necessary through an external cause. It is Averroes' contention that what is in its own nature possible, cannot be converted into the necessary or the eternal, a contention already articulated in *Question* V. Averroes then poses a series of questions to himself. These deal with problems concerning the celestial sphere, its motion, and the source of its motion, and how these problems are to be resolved if it is true that the sphere and its motion are necessary, or eternal, in and of themselves.

1. Still another treatise which also belongs to Averroes.[1]

2. Averroes said: Avicenna was of the opinion that beings are to be divided into three classes:[2] first, beings which are merely possible, that is, the beings which are generated and corrupted; second, beings which are possible in virtue of themselves, necessary in virtue of something other than themselves, and, in his opinion, these are those beings other than the First Principle, which are ungenerated and incorruptible,[3] for in his opinion these latter things only acquire existence, that is to say, necessary existence, from the First Principle; third, a being which is necessary in virtue of itself, and, in his opinion, this is the First Principle alone, and it is this which he calls "necessary of existence."[4]

3. This classification of beings is true; that is, it is true that some beings are generated and corrupted, and that some are eternal on account of causes through which they are eternal, and that some are eternal in virtue of themselves. But the assertion concerning things which are eternal because of eternal causes: that they are necessary in virtue of something other than themselves,[5] possible in virtue of themselves, is not a true characterization;[6] for that which, with reference to its own nature, is possible is not susceptible of eternity --unless some of the possible things can be susceptible of eternity, i.e., unless the nature of the possible can be converted into the eternal.[7] But it had already been demonstrated at the end of the first book of *De Caelo* that the eternal contains no potentiality or possibility at all,[8] and that if it had, that potentiality or possibility would be vain.[9]

4. Someone might say, "Then how do you distinguish between those things which are eternal because they have eternal causes and those things which are eternal in virtue of themselves?" We would answer by maintaining that only the eternal things, in so far as they are eternal themselves, are eternal; and that there are no eternal things in which eternity is something created [in them] in addition to their substance; rather, the fact that they are eternal and that they are beings and essences is one and the same thing.[10] Some of them, however, have their substance, which is described as eternal, because of something other

than themselves; while other have their substance, which is described as eternal, in virtue of themselves.[11]

5. Someone might then say, "This is obvious in the case of the simple, incorporeal beings, that is, the separate Intelligences which are below the First Principle, but how does it apply in the case of the heavens which are compounded of matter and form? It has, indeed, been asserted that every body has a finite force, and that the heavens only acquire an infinite force, and eternity, because of a nature which is eternal in virtue of itself.[12] If this is true, the heavens must be possible in virtue of themselves, necessary in virtue of something other than themselves. However, from what has been demonstrated in the first book [of *De Caelo*] it follows that there is no eternal thing which is possible in virtue of itself. How would you resolve this difficulty?"[13] In response to this objection, we would maintain that the heavens are compounded of matter and form, but that because the heavens have no potentiality, except potentiality with respect to place,[14] as has been shown elsewhere, their matter is something actual, not something potential.[15] For this reason, the use of the term "subject" is more appropriate to them than the use of the term "matter."[16] If the heavens did possess potential matter, they would be generated and corrupted, and they would necessarily have a contrary, and privation would be conjoined to their matter.[17] But it has already been shown that the heavens have no contrary,[18] and that there is no potentiality at all in them at any time,[19] neither as regards the form which exists in them, nor as regards its contrary.[20] This being the case, it follows that the heavens must be necessary with respect to their corporeal substance[21] in virtue of themselves, not in virtue of another cause which bestows on them that necessity which exists with respect to their substance.[22] The heavens, therefore, are not possible with respect to their corporeal substance as is the case with respect to the substance of the generated and corrupted bodies.[23]

6. Again, someone might say, "Since potentiality and possibility do not exist in the heavens except with respect to place alone, their locomotion must be possible in itself, necessary because of their eternal mover.[24] For were it not for the heavens' eternal mover, it would be possible for them to come to a standstill; and if it is possible for them to come to a standstill, they must already have done so an infinite number of times in infinite past time. This being so, it follows that something does exist which is possible in virtue of itself, but necessary in virtue of something other than itself." To this objection we reply that it had already been shown at the beginning of the eighth book of *Physics*[25] that motion, as a genus, cannot fail,[26] so that there might be a time in which there would be nothing in motion. Had a time of this description existed in the past, it would not have been possible for motion to be created [anew] after it, for it is necessary that there be a motion both before and after every one of these generated and corrupted motions.[27] It is also impossible, therefore, for some time to come in the future in which there will be nothing in motion: were this possible, a time of this description would already have existed in the past, and, had this been the case, it would not be possible for anything in motion to exist now, in our time.[28] Because this is so, and because it has been demonstrated that all of these [generated and corrupted]

motions terminate at the motion in the uppermost celestial region,[29] which encompasses all the other motions,[30] it follows that that motion must be necessary of existence in virtue of itself. For if it were possible for this motion in virtue of itself to fail, it would be possible for motion as a genus to fail, and for a time to come in which there would be nothing in motion: inasmuch as if we imagine this motion to be non-existent, we must imagine the rest of the motions to be non-existent,[31] because this motion encompasses and includes the rest of the motions and is the reason whereby motion as a genus does not fail. This being so, it follows that this motion does not require an eternal mover by reason of its being a possible motion; it requires an eternal mover only in so far as every motion must have a mover.[32]

7. The property of this motion which is necessary in virtue of itself, but which requires the mover in so far as it is a motion, is that its mover is not in matter at all; that is, that the mover has no inherence in that which is moved,[33] for every motion which is [caused by an inherent mover] of this description is possible, not necessary.[34] This is because it had already been demonstrated that every force in a body is finite with respect to motivity, and it is with respect to this [motive] force that it is said that every body has a finite force. Consequently, the property of the mover which is not in matter is that it does not cause a motion which is possible in any of its parts; and if this were not the case, the possible could become necessary.[35] The property of the motion which is necessary in virtue of itself is that it is caused by an eternal mover, and that it is not caused by a non-eternal mover, for, in the latter case, a mover which is possible with respect to causing motion would become necessary with respect to causing motion.[36]

8. We have explained this principle which is obscure but which follows from their principles. Many of those who came after Aristotle, who considered themselves to philosophize according to his doctrine, have been in doubt concerning the answer with regard to these matters--so much so that we find that Alexander, in one of the works attributed to him, accepted the proposition which states that the heavens are a body, and that every body has a finite force.[37] And in response to the question of the eternity of the heavens, he said that they only acquire eternity from the Prime Mover. But this assertion is identical with the assertion that there is something possible in virtue of itself, necessary in virtue of something other than itself.[38]

9. Elsewhere, we ourselves have distinguished[39] the number of ways in which it can be said that every body has a finite force, and we have demonstrated which of them is valid with respect to the first body and which is not.[40] We have demonstrated that the term "infinite" is used with respect to the speed of motion, and also with respect to the perpetuity of motion and its continuity.[41] That use of the term which has reference to speed and, in general, being acted upon and acting,[42] is not applicable to any body, be it eternal or not, for it is impossible for anything to be moved in an instant or to act in an instant. But that "infinite" which exists with respect to the continuity of action and being acted upon does exist in the celestial bodies, although it does not exist in what is below them.[43]

10. All these matters are evident to one who possesses perfectly the

principles which the Aristotelians,[44] and those of the reputable[45] philosophers who followed them, employed with respect to these matters.

11. The *Question* has been completed. Praise be to God!

NOTES

INTRODUCTION

1. See, e.g., C.B. Schmitt, "Renaissance Averroism Studied through the Venetian editions of Aristotle and Averroes (with Particular Reference to the Giunta Edition of 1550-52)," in *The Aristotelian Tradition and Renaissance Universities* (London, 1984), No. VIII; H.A. Wolfson, "The Twice-Revealed Averroes," *Speculum* 36 (1961), pp. 373-74, and 383-84.

2. See H.A. Wolfson, "Revised Plan for the Publication of a *Corpus Commentariorum Averrois in Aristotelem*," *Speculum* 37 (1963), pp. 88-94.

3. See, e.g., *LM* XII, c. 13, 299 C-H (Arabic: pp. 1461, 10-1464, 12), and M. Bouyges, *Tafsîr mâ ba'ad aṭ-Ṭabî'a* (Beyrouth, 1938-1948), v. 1, pp. cxvi-cxxxiii.

4. For specific information on which Greek commentaries on the various works were translated, see F.E. Peters, *Aristoteles Arabus* (Leiden, 1968). It is not clear if Ibn Rushd's knowledge of John Philoponus stems from the latter's commentaries, a number of which had been translated, or simply from the refutations of Philoponus written by later scholars, such as al-Farabi. The commentaries of Simplicius to *Physics* and *De Caelo* are not documented as having been translated, but see below, *Question* II, note 1.

5. I.e., the exponents of the Kalâm, a system of rational argumentation designed to support the tenets of Islam.

6. See M.J. Müller, *Die Philosophie und Theologie von Averroës* (München, 1875). *Kitâb al-Kašf* also did not receive a Latin translation.

7. But see below, p. xvi.

8. See the remarks of A. Hyman, *Averroes' De Substantia Orbis* (Cambridge, Mass. and Jerusalem; 1986), pp. 15-16.

9. See E. Renan, *Averroès et l'averroïsme* (Paris, 1861), pp. 456, 458, 463; M. Steinschneider, *Die Hebraeischen Uebersetzungen des Mittelalters und die Juden als Dolmetscher* (Berlin, 1893), pp. 96-97.

10. Narboni = Moses ben Joshua of Narbonne, a fourteenth century philosopher, and devoted follower of Averroes.

11. The treatises most often found are: Abu al-Qasim ibn Idris, *On Prime Matter*; *Idem*, *On the Forms*; and Abu Dja'afar ibn Sabaq, *Treatise on the Prime Matter*. For a listing of the works by other authors and other works by Averroes contained in the MSS, see Steinschneider, *op. cit.*, pp. 188-89 *et passim*, and S. Rosenberg, "The Hebrew Translations of Averroes' *Questions in Physics* and the Commentary of Moses Narboni,"

Qiryat Sêfer 57 (1982), pp. 719-21 [in Hebrew]. Rosenberg also indicates the distribution of these works in the various MSS.

12. MS. Munich 36, and according to Steinschneider (*op. cit.*, p. 182), MS. Leipzig 40. On this excerpt, see H. Tunik Goldstein, "New Hebrew Manuscript Sources for Averroean Texts," *JNES* 38 (1979), p. 31.

13. The most obvious example of this variation is found in the title of the work: the term for "question" is not *derûš*, as in the "standard" version, but *še'êlâ*. In the "standard" version, each treatise after the first is designated as *ma'amâr*; here, the term is, again, *še'êlâ*. Apart from this, the variants may simply be peculiarities of the scribe, as are his spelling variants. I have seen only one of the MSS belonging to this group.

14. On this work and its translator, see Goldstein, *op. cit*, pp. 29-31. I have since seen two more exemplars of this text.

15. But see Steinschneider, *op. cit.*, p. 178. Rosenberg (*op. cit.*, p. 716) states there can be no doubt that Narboni contributed to ordering the *Questions*, and that Narboni himself testifies to this in his commentary to *De Substantia Orbis*. Unfortunately, he cites no reference. But Hyman, (*op. cit.*, pp. 15-16), apparently takes Narboni's remarks to apply only to *De Substantia Orbis*. (See also *Idem*, "The Composition and Transmission of Averroes' *Ma'amar be-'Eṣem ha-Galgal*," *Studies and Essays in Honor of Abraham A. Neuman* [Leiden, 1962], pp. 299-307. There, Hyman sees Narboni's work as based on an earlier recension or recensions.) Certainly, Narboni makes no mention of any role he may have had in collecting or arranging the treatises of the *Questions* in his preface to the work, at least in the sole version I have seen (MS. Paris, 988).

16. See Steinschneider, *op. cit.*, p. 178, especially n. 534.

17. *HU*, p. 96

18. *Questions* I and VI; it might be argued that *Question* VII also falls into this category.

19. But cf. Steinschneider, *op. cit.*, p. 96, n. 363. Our *Questions* are not like the *Ḍamîma* (which Averroes says he wrote in response to a request), nor are they like Avicenna's *Kitâb al-mabâḇi<u>th</u>ât*, a report of actual conversations in which questions were put to Avicenna to which he responded. Rather, these works are firmly centered in the tradition of commentary and investigation exemplified by the Greek commentaries.

20. The last named work often accompanies the *Questions* in the manuscripts, and is sometimes subsumed under the title *Questions in Physics*.

21. *Question* VIII

22. *Question* VII, however, most probably is an appendix to

LP VIII. Note the reference in *EP*, cited by Steinschneider (*op. cit.*, p. 180). This probably accounts for the citation of this work as *ta'alîq* in the Arabic bibliographers (see E. Renan, *op. cit.*, p. 464). But see the description of the Arabic MS, below.

23. Cf. *QLog, Libri priorum*, 100F.

24. *Question* VII; see below, the description of the Arabic MS.

25. *Questions* I, VI, VII, and VIII

26. See below, *Question* IX, n. 1.

27. Narboni, in his Introduction, states:

> After we had completed our commentary to Ibn Rushd's Middle Commentary on the *Physics*, we saw fit to comment on the *Questions* of Ibn Rushd concerning general matters related to that book, *Questions* which are of great value and precious in knowledge...

28. Of the six other Hebrew MSS listed by Steinschneider (*op. cit.*, p. 179), the Turin MS has long been gone; I was not able to locate the Berlin MS, nor the Pinsker; and I was unable to obtain copies of the remaining MSS. I was ignorant of the existence of MS. Paris 1341 until the article of Rosenberg (*op. cit.*, p. 715) appeared. On that MS, see the brief, and confusing, remarks of M. Schwab, ""Manuscrits du supplément hébreu de la Bibliothéque (*sic*) Nationale," *Revue des études juives* 37 (1898), p. 130.

29. In the case of *Question* VII, because of its length, the commentary interrupts the text.

30. See *Bibliothecae Apostolicae Vaticanae codicum manuscriptorum Catalogus* (St. Ev. Assemanus et Jos. Sim. Assemanus recens.; Rome, 1756), I, i, pp. 435-36. See also the description in Hyman (*Substance*, p. 20).

31. See M. Steinschneider, *Die Hebraeischen HSS der K. Hof-und Staatsbibliothek in Muenchen* (Muenchen, 1875), pp. 15-16, and also pp. 22-23. See also Hyman, *op. cit.*, pp. 20-21.

32. See the general description in Hyman, *op.cit.*, p. 23.

33. See Steinschneider, *HU*, p. 179.

34. See below, *Question* VIII, n. 34.

35. Treating only *DSO*, Hyman (*loc. cit.*) speaks of annotations in four different hands, but in the text of the *Questions* the marginal annotations appear to me to be in the hand of the original scribe (with the exception of Steinschneider's notes).

36. See H. Zotenberg, *Catalogues des manuscrits hébreux...de la Bibliothèque Impériale* (Paris, 1866), I, i, p. 176.

37. Here called *še'êlâ*; cf. n. 13, above.

38. See Steinschneider, *Hebraeischen HSS...in Muenchen*, pp. 19-24.

39. See *Catalogus librorum manuscriptorum qui in Bibliotheca senatoria civitatis lipsiensis asservantur* (Codices orientalium linguarum descripserunt H.O. Fleischer et F. Delitzsch; Grimae, 1838), pp. 304-05.

40. See Steinschneider, *HU*, pp. 178-79.

41. Steinschneider (*op. cit.*, p. 180) errs: it is not *Question* IV that is missing in this MS, but *Question* III.

42. See Goldstein, *op. cit.*, p. 31.

43. See below, *Question* III, n. 52.

44. See Zotenberg, *op. cit.*, pp. 176-77.

45. See Goldstein, *op. cit.*, p. 29.

46. See below, *Question* IX, n. 1.

47. Zotenberg, *op. cit.*, pp. 184-185

48. *Catalogue of the Hebrew and Arabic Manuscripts in the British Museum* (ed. G. Margoliouth; London, 1965), III, pp. 188-89; the entire MS is described on p. 186.

49. See below, *Question* VII, n. 56.

50. On Ṭodros Ṭodrosi's diction, see G. Vajda, "Les deux versions hébraïques de la dissertation d'Averroès sur la science divine," *Revue des études juives*, N.S. 13 (1954), pp. 64-66.

51. See H. Derenbourg, *Les Manuscrits arabes de l'Escurial* (Paris, 1884), I. p. 440. Earlier this MS was known as Escurial 629 (see M. Casiri, *Bibliotheca Arabico-Hispana Escurialensis* [Madrid, 1740], I, p. 184). Steinschneider (*HU*, p. 180) incorrectly refers to this MS as Esc. 632.

52. On the order known to Narboni, see Steinschneider, *op. cit.*, p. 181, n. 546.

53. I am deeply indebted to Professor William Thomson for his assistance in deciphering this text.

54. Specific instances are cited in the notes to *Questions* VI and VII.

55. The most common grammatical failing in the Hebrew MSS is an utter disregard for gender in pronominal suffixes. As might be expected, the spelling of the MSS is wildly inconsistent as regards *plene* and *defectiva*. I have not attempted consistency, or correction, with one exception: a consistent spelling of Aristotle's name.

56. Indications of punctuation occur regularly only in the Vatican MS, often in very surprising places. In MS. Mun. 36, "paragraphing" is indicated by a horizontal line over the appropriate word, but the marks are few and far between.

57. But see M. Steinschneider, "Alfarabi," *Mémoires de l'academie impériale des sciences de St.-Petersbourg*, VIIe Série, XIII (1869), p. 150. I, however, cannot find any references to the *Questions* in *LP* VIII, Comm. 1 (although this commentary has many points of similarity with *Question* VII); and the reference in *LP* VIII, Comm. 78, 424 M, to an explanation *in nostris quaestionibus*, most probably means *QLog*. See below, *Question* VIII, nn. 47 and 48.

58. For an indication that Averroes' version of *Physics* was not identical with the translation preserved in the (unique) Leiden MS, see below, *Question* III, n. 9, and *Question* VII, n. 114.

59. It should be noted that, through some error, Rosenberg's table of the contents of the MSS and their order (*op. cit.*, p. 721) omits *Question* V from the contents of the Vatican MS. The cursory description of S. Munk, (*Mélanges de philosophie juive et arabe* [Paris, 1927], pp. 436-37; 438, no. 6; 503-04) is based solely on MS. Paris 988, which is itself described inaccurately. Renan (*op. cit.*, pp. 65-73), in his list of Averroes' works, does not even mention *Question* VII, even though that work is listed in the Arabic bibliographic sources he cites.

60. See below, *Question* III, n. 1, and *Question* IX, n. 1.

61. See especially pp. 121-22, but note also pp. 112 and 162.

62. In *Avicenna: Scientist and Philosopher* (ed. G. Wickens; London, 1952), pp. 46-48; cf. *Idem*, "Averroès inconnu," *Actes du XXe Congrès International des Orientalistes* (1938), pp. 337-38.

63. See below, *Question* V, n. 2.

64. *The Philosophy of Abraham Shalom* (Berkeley and Los Angeles, 1964), pp. 19-26.

65. *Proofs for Eternity, Creation and the Existence of God in Medieval Islamic and Jewish Philosophy* (New York and Oxford, 1987), pp. 311-335

66. "John Philoponus as a Source of Medieval Islamic and Jewish Proofs of Creation," *JAOS* 89 (1969), pp. 359-60; unfortunately, the text is quoted from Steinschneider's translation of a single MS.

67. *Avicenna and the Aristotelian Tradition* (Leiden, 1988), pp. 118-19

68. See his commentary to *Moreh Nebukim* I, 74, and II, Introduction, Proposition XV. For Narboni's opinion of the *Questions*, see above, n. 26.

69. *Mip'âlôt 'Elôhîm* (Lemberg, 1863), II, 1, p. 12b; Steinschneider (*HU*, p. 179) gives the reference, incorrectly, as III, 1.

70. P. 22, ll. 13-14, and l. 24, in the edition of Steinschneider (*'Ôsar Tôb* VII [1881]).

71. The passage in question is cited below, *Question* III, n. 51.

72. "Alfarabi," p. 162; see also *HU*, p. 204, n. 705.

73. Steinschneider's further conjecture that Elijah del Medigo (Elias Cretensis; d. 1496) was acquainted with these *Questions* (*HU*, p. 183, n. 562; *Hebraeische Bibliographie* XXI [1881/82], pp. 64 f.) rests solely on del Medigo's knowledge of the names of the Muslim authors whose works appear together with the *Questions* in a large number of MSS. While it seems inherently probable that del Medigo did know our text--he wrote both a Hebrew and a Latin commentary on *De Substantia Orbis*, which, as we have seen, appears in many of the same MSS as the *Questions*--, he does not quote from, or refer to, our text in his *In dictis Averrois super libros physicorum clarissimae adnotationes* (Venice, 1551). There, his quotations of Averroes are taken from *EP* and *IP*.

74. *Isaac Albalag*, (*Études de philosophie mediévale* 49 [Paris, 1960]), p. 135, n. 2

75. See the text published by Osias Schorr, *He-Ḥâlûṣ* VI (1861), p. 87.

76. *Abraham Shalom*, p. 14

77. *Op. cit.*, pp. 51-52

78. *Liber de mundi creatione physicis rationibus probata* (Venice, 1527); on this work, see P. Duhem, *Le système du monde* (Paris, 1913-1917), IV, p. 497; V, p. 216.

79. It is not clear that this MS was owned by Egidio. It does not appear, at least by title, in the catalogue of his library published by C. Astruc and J. Monfrin, "Livres latins et hébreux du Cardinal Gilles de Viterbe," *Bibliothèque d'humanisme et renaissance* 23 (1961), pp. 551-554.

80. But see below, *Question* II, and *Question* IX.

81. *Question* IX is a good example of this.

82. Munk, *Mélanges*, pp. 503-504; Steinschneider, *HU*, p. 178

83. This, however, would not explain why, for example, Abraham de Balmes in translating *De Substantia Orbis VII* (from the version of Todros Todrosi) and attaching it to the body of that text, did not do the same thing with our *Question* IX, which immediately precedes it in the MSS of that translation, and which is closely related in subject matter. If one hypothesizes that de Balmes was aware that *Question* IX belonged to another collection, one may still wonder why he found that collection unworthy of translation.

84. The circular view of motion and time is vigorously upheld in *Question* III. It is also applied to the procession of created things.

85. This forms the main argument of *Question* IV.

86. This argument occurs at the end of both *Question* VII and *Question* IX.

87. This response is most clearly seen in *Questions* V and IX. It is largely on the ground of his objection to the notion of the possible acquiring necessity, or eternity, that Averroes frequently lumps John Philoponus and Avicenna together in polemical passages, and (correctly) attributes their views to Plato, as well.

88. *Question* V

89. *Question* IX; this contradicts the view expressed in *Question* VII (and elsewhere) that the motion of the sphere is eternal because its mover is eternal.

90. See, e.g., *Physics* VII, 1, 242b, 32-243a, 2 (in the *textus alter*).

91. See *LP* VIII, Comm. 78, 424 K-M.

92. This is the burden of *Question* VIII, although references to the problem and its solution are also found elsewhere.

QUESTION I

1. This *Question* cannot be identified with any of the specific titles of Averroes' works listed in the Arabic bibliographies. It could, of course, be one of the works referred to under the blanket title "Questions in Philosophy" (see Renan, *Averroès*, App. III, p. 456; App. IV, p. 458).

2. Aristotle's name does not appear in the Hebrew text. As in many mediaeval philosophical works, he is simply called "the philosopher."

3. See below, n. 16.

4. On the fact that this proposition is not self-evident, see *TAT* I, p. 2 (Arabic: p. 5, 10-15; Latin: 15 GH), and *ibid.*, XIV, p. 287 (Arabic: p. 417, 5-12; Latin: 117 BC).

For another interpretation of *Physics* VII, 1, 241b, 24-242a, 15, see below, *Question* VII, paragraphs 6-8.

5. Averroes defines being moved primarily--negatively--at *IP* VII, i (fol. 101a, 15-23):

> What is not moved primarily is what is moved in its entirety because some part of it is moved *per se*: such as the motion of a certain magnitude of earth which is greater than that magnitude which is the smallest part of earth [because of the motion of the smallest part], and such as the locomotion of the animal because of the motion of the elemental heat in it. It is obvious that what is not moved primarily is moved by something else: namely, by what is moved primarily. It is [therefore] clear that if something exists which moves itself so that the mover [in it] and what is moved are one and the same thing, this must be moved primarily.

In *LP* V, Comm. 3, 207 M-208 A, Averroes defines that which is moved *per se* as that which is not moved because of another mover, while that which is moved primarily is defined as that which is not moved because another part of it is moved primarily. He also notes that sometimes the terms "*per se*" and "primarily" are synonymous.

In this *Question*, "motion *per se*" is not to be opposed to "accidental motion," i.e., the motion of a contained part together with the whole, or the motion of a quality together with the motion of its subject. Rather, "being moved *per se*" stands in opposition to "being moved by a part." (For the distinctions involved, see *Physics* V, 1, 224a, 21-29.) Similarly, at *LP* VII, Comm. 2, 307 GH, "Galen and others" are reproached for failing to understand Aristotle's proof at the beginning of the seventh book, their failure being due to confusing accidental motion with the motion caused by a part of what is moved.

It should, however, be noted that in *LP* VII, Comm. 1, 306 F, Averroes himself identifies "being moved by a part" with accidental motion.

It is also possible to find passages in which Averroes opposes motion *per se* to both accidental motion and motion through the motion of a part, e.g., *IP* V, i (fol. 70b, 21-71a, 5):

> We say that the altered and the moved, as well as the cause of alteration and the cause of motion, are each said to be of three kinds: essential [*per se*], accidental, or through a part.... What is moved *per se* is what is not moved either accidentally or through the motion of some part of itself...

Averroes' definition of "*per se*" in this *Question* agrees with the comment of Yaḥya ibn ʻAdi to *Physics* VII, 1 in the (unique) Leyden MS of the Arabic *Physics* (p. 743, 7-8 [all citations of the Arabic text of

Physics, and the accompanying commentaries by Yaḥya ibn ʻAdi and Abu'l-Faraj, refer to the edition of A.Badawi, *al-Ṭabîʻa* (Cairo, 1964-65)]; on the MS itself, see S. Stern, "Ibn al-Samḥ," *Journal of the Royal Asiatic Society* [April, 1956], pp. 31-44). But both Alexander of Aphrodisias in his "Refutation of Galen," (67b, 14 ff.; cf. 67a, 6-8), a work which appears to be an excursus excerpted from his commentary on *Physics* VII, 1, and Simplicius (*in Aristotelis Physicorum Libros Quattuor Posteriores Commentaria*, *CAG* X [ed. H. Diels; 1895], p. 1037, 20-22; cf. *ibid.*, 25-26) do identify motion because of a part with accidental motion (see below, note 24).

On the "Refutation of Galen," see F. Rosenthal, "From Arabic Books and Manuscripts V: A One-Volume Library of Arabic Philosophical and Scientific Texts in Istanbul," *Journal of the American Oriental Society* 75 (1955), pp. 14-23; and S. Pines, "*Omne quod movetur necesse est ab aliquo moveri*: A Refutation of Galen by Alexander of Aphrodisias and the Theory of Motion," *Isis* 52 (1961), p. 22. I am greatly indebted to Professor Rosenthal for a photostat of the MS.

6. The Hebrew text reads only "primariness." Cf. the general definition of "primariness" in *Physics* VI, 5, 235b, 33-34. Here, the specific reference is to *Physics* V, 1, 224a, 26-29; cf. *ibid.*, VII, 1, 241b, 38.

Other treatments of "what is moved primarily" in Averroes' commentaries include:

IP VII, i, fol. 101a, 12-15:

> That which is moved primarily is that whose motion does not take place because some part of it is moved *per se*, or can be moved *per se*, such as the motion of the smallest part of earth or water, i.e. a part such that a smaller magnitude cannot receive the form of water; for in the natural bodies [elements] this [minimum] magnitude is fixed.

EP VII (Hebrew: p. 35a, 13-21; Arabic: p. 114, 3-9):

> That which is moved primarily, i.e. that which is not moved because of some part in it, is that which is moved *per se*. In the case of these simple bodies, with respect to which some question has occurred, it is the smallest quantity [e.g.,] of fire which can be moved upward, or the smallest quantity of earth which can be moved downward. Any part of earth or fire answering to this description which undergoes motion is that which is moved primarily, for to no part of that part can that motion belong, inasmuch as fire [for example] has no smaller part. This is so only because the [minimum] magnitudes of existing things are fixed.

7. Cf. Simplicius, *in Physicorum* VII, 1 (p. 1037, 27-28): "But first he sets straight our thinking...."

8. What follows is a direct quotation of *Physics* VII, 1, 241b, 29-34 (28-33 in the *textus alter*), but Aristotle says that AB is "moved by itself," whereas our text reads that AB **is not moved by anything [else].**

As Simplicius (*op. cit.*, p. 1036, 4-6) noted, *Physics* VII, 1(-3) exists in two versions, "having only some few verbal differences; for the problems are the same, and the demonstrations of them occur in the same order in both versions." The version of the text which is designated as *alpha* has generally been considered to be the better (W.D. Ross, *Aristotle's Physics* [Oxford, 1936], Commentary, p. 14).

Averroes does not seem to have been aware that there were two versions of the text. On the basis of the quotations and allusions in this *Question*, it would appear that he was acquainted with a text representing a conflation of the two versions. In fact, the text as quoted in this *Question* is very close to the text of the Arabic *Physics* as found in the Leyden MS (p. 734, 1-6), especially if the editor's unnecessary emendation (line 4) is omitted. It is Badawi's contention (Introduction, p. 7) that the Arabic version is based on the *textus alter*, but I find, particularly in the text quoted here, a strong influence from *alpha*. (Cf. the Latin translations of Aristotle printed in *LP*.)

9. Three of the MSS (the fourth MS omits) read the equivalent of EF, and this is plainly incorrect. I conjecture that the Hebrew text, like the Greek, originally used three letters, the equivalents of DEF, and that at some point the first letter was dropped.

10. That is to say, it cannot be known by sense perception. The Arabic text of *Physics* VII, 1, unlike the Greek text, does not specify "sense perception." But cf. our text, below, paragraph 5.

11. At first glance, the tentative way in which Averroes puts forth this opinion seems strange. It is to be accounted for, however, by the difference in reading between the Greek text of *Physics* and the Arabic version (see above, n. 10).

12. **Differentiation**: the Hebrew term used here is *šinnûy*, used in the same sense as *ḥillûp*, the term that will be used later in this text. *šinnûy* is used as an equivalent for *ḥillûp* when the latter has the meaning "change" (*metabolê*; see H.A. Wolfson, *Crescas' Critique of Aristotle* [Cambridge, Mass.; 1929], p. 501). Our text would seem to indicate that *šinnûy* took on all the meanings of *ḥillûp*.

13. Averroes' language here is drawn from two places in *Physics* VII: **the source of its motion** is taken from ch. 2, 243a, 32-34 (3-4 in the *textus alter*), where the mover and that moved are said to be "together"; **which is continuous with that thing** is taken from ch. 1, 242b, 59-63 (24-27 in the *textus alter*), where it is said that the mover and what is moved must be continuous or in contact with one another. Both these sources are passages in which Aristotle discusses a series of bodies in contact in order to show that the series must end with an unmoved mover. Cf. *EP* VII (Hebrew: pp. 36a, 28-36b, 1; Arabic: p. 117, 10-12):

> ...then it is clear that they form a magnitude which is one by contiguity, and that their motion is simultaneous, as is the motion of the parts of one continuous magnitude.

14. Two MSS read DE. See above, n. 9

15. *Laws* X, 894B-896B; *Phaedrus*, 245C-E. It is curious that Averroes should apply Plato's conception of the self-moved to a body, but see Pines, *op. cit.*, p. 33 and n. 60.

This is the only Averroean text, commenting directly on *Physics* VII, 1, where it is conjectured that Aristotle's argument was aimed at Plato, but such a claim is made in *LDC* III, Comm. 28, 198 LM, and *ibid.*, IV, Comm. 22, 249 FG. In both these texts it is also said that the argument of this chapter refers to the motion of the elements, a statement missing in this *Question* (see, however, *Question* VII, paragraphs 6 and 23, and n. 22; see also the quotation from *EP*, n. 6 above). Alexander, too, referred this argument to the motion of the elements (see Pines, *op. cit.*, pp. 27 ff.). Here, again, this *Question* contradicts Alexander's "Refutation of Galen": Alexander claims that Plato, too, held that everything in motion has a mover. See Pines, *op. cit.*, pp. 24-25.

Averroes' other references to Plato's self-moved, while frequent, have nothing to do with establishing the purpose of *Physics* VII, 1: e.g., *LP* II, Comm. 3, 49 B-D; *ibid.*, VIII, Comm. 40, 380 A-F, and Comm. 76, 422 L; *EP* VIII, p. 142, 3-5 (Arabic).

16. **At the opening [of this book]**: Perhaps one ought to translate this as "in this proem," for according to *LP* VII, Comm. 1, 306 HI, Aristotle's argument up to this point constitutes only a preliminary argument designed to show that the evidence of sense perception is not sufficient to establish the existence of a self-mover, i.e. one in which the mover and the moved are identical. Cf. Simplicius, *op. cit.*, p. 1038, 12-19.

The remainder of our *Question* consists of an analysis of the logical form of *Physics* VII, 1, 241b, 44-242a, 49 (241b, 34-242a, 15 in the *textus alter*), ending with the reconstruction of Aristotle's argument in a syllogism of the first figure. In the "Refutation of Galen," Alexander is also concerned with the logic of the argument, and his analysis also culminates in a reconstruction of the argument in a syllogism of the first figure (see Pines, *op. cit.*, pp. 29 ff.). Alexander, however, is not so much concerned with the form of Aristotle's argument, but with the logical validity of the propositions involved. Averroes here appears to take the validity of the propositions for granted; indeed, he says that the first of them is self-evident (paragraph 7). He also appears to consider the argument demonstrative, while Alexander characterizes the argument as "logical" (see below, n. 26; Pines, *loc. cit.*).

In other writings, Averroes is concerned with both the validity of the propositions and the worth of the argument (see *EP* VII, pp. 115, 17-116, 11 [Arabic; quoted below, *Question* VIII, n. 56]; *LP* VII, Comm. 2, 307 G-I). In neither of these texts is Alexander mentioned, although in the Introduction to *LP* (1 B and E) Averroes states explicitly that, even though he did not have all of Alexander's commentary on *Physics*, he did have the commentary to Book VII (cf. Steinschneider, *HU* p. 123, n. 104).

It might also be noted that in *EP* VII, Averroes does not discuss the "proem," but prefaces the treatment of the propositions with mention of the motion of the elements. Alexander, too, does not discuss the "proem." His discussion of Book VII begins with the propositions (at fol. 67a, 25), the earlier part of the treatise consisting of polemic and a summary of the way in which the proof that everything in motion has a mover is treated in Book VIII.

At the end of his discussion of the propositions in *IP* VII, i, fol. 102a, 13-102b, 3, however, Averroes writes:

> Great confusion had already occurred to some men about this proof, so that for some time even we ourselves were confused about it... and the reason for this is that we did not comprehend Aristotle's stipulation with respect to that thing in motion which is assumed to be moved primarily... and, moreover, that no explanation of the demonstration by the commentators had fallen into our hands; for Themistius omitted it from his book, and Alexander's commentary was also missing from his book in the translation which had come down to us. Later we found the commentary of Alexander on this passage and found that it agreed with what we had written, even though it is lengthier in explanation. [But] it may, perhaps, also be more profound.

It is not clear from this text whether Averroes refers to the whole of Alexander's commentary on *Physics* VII, or only to the excursus now called the "Refutation of Galen." It *is* clear that finding Alexander's text was of some significance for his own position, and that he considered that he was in perfect agreement with Alexander (but see Pines, *op. cit.*, p. 33, n. 59). It may well be that our *Question* belongs to an early stage of Averroes' career, prior to his finding Alexander's commentary.

17. This is Aristotle's wording in the *textus alter* (242a, 1). Alexander (*op. cit.*, fol. 67b, 21-23) and Simplicius (*op. cit.*, p. 1038, 27-31), apparently relying on *alpha*, give this proposition in the form: "What is not set in motion by something else will not necessarily cease...." They also insist on the addition of "necessarily" in the inverted proposition (see below, paragraph 7): "What necessarily comes to rest...." In the commentary of Abu'l-Faraj to the Arabic text of *Physics* (p. 740, 15-17), the terms "necessarily" and "not ever" do not occur at all (note also the inversion in the order of the propositions). Yaḥya (p. 743, 2-3) uses "necessarily " in the original proposition, but not in the inverted proposition.

18. See Wolfson, *Crescas*, p. 541. Reference to some kind of conversion or inversion seems to be a commonplace in commentaries on this argument, but it is not found in the "Refutation of Galen" (unless we emend the text at fol. 68a, 13) or in *EP* VII, pp. 113, 5-117, 9 (Arabic) which has many points of similarity with Alexander's work. See, however, Simplicius, *op. cit.*, p. 1040, 21-22; Abu'l-Faraj, p. 740, 18; Yaḥya, p. 743, 4-5; *LP* VII, Comm. 2, 307 C; and below, *Question* VII, paragraph 6.

19. This type of demonstration, its form, and its validity is treated at some length in *Questions* VII and VIII and in the notes to those *Questions*. On Aristotle's procedure in *Physics* VII, cf.Simplicius, *op.*

cit., p. 1039, 1, and *LP* VII, Comm. 2, 308 AB.

20. See *Physics* VI, 10, 240b, 8-241a, 26; *ibid.*, VI, 4, 234b, 10-20; and *EP* VII (Hebrew: p. 35a, 6-9; Arabic: p. 113, 9-11):

> With reference to the first proposition [that everything in motion *per se* and primarily is divisible], he employs in its demonstration what has already been demonstrated in the sixth book: that everything in motion is divisible into that "from which" and that "to which," and [also] what was demonstrated there [to the effect] that what is not divisible cannot be moved.

21. The Hebrew term used here for **necessarily** (*be-hekrêaḥ*) can also have the meaning of "by compulsion." Although the latter meaning is clearly inappropriate to our text, the term is used in that sense at *EP* VII (Hebrew: p. 35a, 22-24; Arabic: p. 114, 10-11): "It is clear that if a part of it is imagined to be at rest, the remainder must come to rest by nature, not by compulsion."

Elias Cretensis [Elijah de Medigo] appends a note to his discussion of these lines from *EP* (*Super libros physicorum*, fol. 154, col. 1 E). The note tells us that in certain writings, Averroes speaks not of "compulsion," but of "necessity." It is del Medigo's opinion that both versions amount to the same thing: for what does not come to rest because of an external impediment, must be at rest because of its own nature; and what is at rest necessarily, cannot be moved any longer:

> In quadam scriptura invenitur, quiescit naturaliter non de necessitate: et ut videtur mihi, quasi in idem redit. nam per hoc intendit Commentator dicere, quod quiescit non propter impedimentum aliquod extrinsecum necessitans ipsum non moveri, sed dimissum suae naturae non impedimentum ab extrinseco non movetur. Cum autem in alia scriptura dicit, quiescit de necessitate, intelligitur qui non potest amplius moveri.

22. Cf. *LP* VII, Comm. 2, 307 G, and, below, *Question* VII, paragraphs 6 and 7.

23. Cf. below, *Question* VIII, paragraph 12.

24. *IP* VII, i, fol. 101b, 16-18:

> Thus, of necessity, it follows when CB, which is the part, comes to rest, that ACB, which is the whole, comes to rest. And when a certain ACB in motion comes to rest due to the fact that its part comes to rest, it necessarily comes to rest through the coming to rest of something other than itself, for the whole is not the part.

The stipulation that the whole is not the part is not found in Averroes' other treatments of this passage. It is found in the "Refutation of Galen" (fol. 67b, 6-7 and 26-27). For our text, the commentary of

Narboni, *ad loc.*, supplies the missing stipulation:

> When Averroes says, **what is moved primarily must be set in motion by some other thing**, he means it must be set in motion by a mover which is not identical with its whole.

25. On the phrase, **it must necessarily be moved by some principle [of motion] in it,** see *Physics* VII, 1, 241b, 35 (25 in the *textus alter*), and the commentary of Yaḥya *ad loc.* (p. 743, 7-8).

26. Cf. *LP* VII, Comm. 2, 307 IK. Aristotle (242a, 45-46 [13-14] omits this intermediate conclusion and immediately concludes that everything in motion has a mover. Averroes does not go on to this final step here.

Alexander (*op. cit.*, fol. 67b, 21-23) also summed up Aristotle's argument in a syllogism of the first figure. Simplicius, however, criticized Alexander's use of the first figure, and offered his own summation in the second figure (*op. cit.*, p. 1041, 10-17).

As we have said (n. 16, above), Averroes offers no characterization of Aristotle's argument in this *Question*. In this he differs from Alexander (n. 16, above), Simplicius (who, agreeing with Alexander, characterizes the arguments of *Physics* VII as "dialectical" [*op. cit.*, p. 1036, 12-13]), Yaḥya (who says that this argument is dialectical, not demonstrative [p. 743, 11-12]), and Abu'l-Faraj (who agrees the argument is dialectical, whereas the arguments of Bk. VIII are demonstrative [p. 740, 2-4]). However, in those commentaries where Averroes discusses the objections that have been raised against Aristotle's argument, he also offers characterizations of that argument. In *EP* VII, p. 116, 7-12 (Arabic), he states that the argument is of the same class as the arguments frequently used in mathematics, where no absurdity results from their use; and of the same type as the arguments in *Physics* VI, where Aristotle assumed that for every moved thing, there is something moved more rapidly and something moved less rapidly (cf. below, *Question* VIII, paragraph 11). In *LP* VII, Comm. 2, 307 KL, Averroes states that even if this argument is not absolutely demonstrative, it is nonetheless true, and belongs to the category of "infallible signs" (see *Rhetoric* I, 2, 1357b, 1-21; cf. Wolfson, *op. cit.*, p. 326; see also, below, *Question* II, paragraph 7 and n. 19).

27. Narboni comments, *ad loc.*:

> When Averroes says that **whatever is in motion must be set in motion by something [else]**, he means everything in motion in place *per se* about which there is doubt as to whether it is moved primarily or whether it contains something moved primarily.

28. **Praise be to God**: this phrase is given a two letter abbreviation here. In medieval Hebrew, this abbreviation would normally mean "Thanks be to God"; but in light of the Arabic text of the ending of *Question* VI, "praise" seems more likely than "thanks." Both words begin with the same letter of the Hebrew alphabet. Cf. *Question* VII, paragraph 49.

QUESTION II

1. Each of the subsequent *Questions* will also begin with this phrase. Steinschneider (*HU*, p. 179) conjectured that this work might be identical with one of the two treatises on the motion of the heavens mentioned in the Escurial list, or with the treatise on that topic cited by Ibn Abi Uṣaybiʻa (see Renan, *Averroès*, pp. 456 and 464). It had, however, also been conjectured that the titles in the bibliographies referred to *De Substantia Orbis* or to astronomical works. See Steinschneider, *op. cit.*, p. 182, n. 553, and Renan, *op. cit.*, p. 75.
This *Question* is a peculiar work. While many of its parts have parallels in other writings by Averroes, the overall plan of the *Question* and its purpose (not revealed until the very end of the work) have no parallel. In addition, we find terminology which is not used, in the sense intended here, elsewhere in the Averroean corpus. It is my belief, which I hope to document fully in a future article, that this *Question* is based on a Greek prototype: there are striking similarities between this treatise and material contained in the commentaries of Simplicius (*in Aristotelis De Caelo Commentaria*, *CAG* VII [1894; ed. I. Heiberg], I, 2, esp. pp. 38, 8 ff.; cf. *Idem*, *in Physicorum* VIII, 9, pp. 1313-1314).
In Averroes' commentaries, the closest parallels to the plan of this *Question* are to be found in *LP* VIII, Comm. 75, 419 D-M, and *LDC* II, Comm. 22 and 23, 109 H-110 K (where the argument deals with the perfection of the circle and the sphere.)

2. The reduction of Aristotle's arguments to neat, syllogistic form, with every premise made explicit, is common in the Greek commentators and very characteristic of Averroes' commentary (see, e.g., above, *Question* I, paragraph 9 and n. 26). The reason for the practice is clear: Aristotle, despite his insistence on the necessity for demonstration, does not do it. See J. Barnes, "Aristotle's Theory of Demonstration," *Articles on Aristotle* 1 (London, 1975), especially p. 66.
The texts which provide the premises for this syllogism are *Physics* VIII, 9, 265a, 17 ff., and *De Caelo* II, 4, 286b, 18-23.

3. At this point, the Hebrew text changes to a new word for **perfection/perfect** (*tammîm* rather than *šelêmût/šelêmâ*). It is likely that this reflects a change of term in the Arabic original. In this text, certainly, the two terms are used synonymously. See, however, S. Pines, "An Arabic Summary of a Lost Work of John Philoponus," *Israel Oriental Studies* II (1972), p. 325, and R. Walzer, "New Light on the Arabic Translations of Aristotle," *Greek into Arabic: Essays on Islamic Philosophy* (Cambridge, Mass., 1962), pp. 95-96.

4. Aristotle defines the perfect, or complete, in (i) *Physics* III, 6, 207a, 8 ff.; (ii) *De Caelo* I, 1, 268b, 4 f.; (iii) *ibid.*, II, 4, 286b, 18 ff.; (iv) *Metaphysics* V, 16, 1021b, 12-1022a, 2; (v) *ibid.*, X, 4, 1055a, 10 ff. Clearly, it is the language of (iii) which has influenced Averroes' wording of the definition here. Other instances of this version of the definition in his commentaries are : *EM* I (Arabic: p. 29, 4-6; Latin: 362 DE; German: pp. 22-23 [van den Bergh]); *LDC* I, Comm.

4, 4 D; *ibid.*, II, Comm. 23, 110 G.
It should be noted that Aristotle himself (*Physics* VIII, 9, 265a, 18 ff.) prefers to emphasize the imperfection of rectilinear motion. Cf. *LP* VIII, Comm. 75, 419 G.

5. Cf. *LP* V, Comm. 42, 233 EF.

6. *Topics* I, 8, 103b, 7-12; *ibid.*, II, 1, 109a, 13-17; *ITp* I, 7, 12 H (Arabic: p. 41, 12-15 Butterworth/Haridi). Cf. *LP* VIII, Comm. 27, 365 D; *Intermediate Isagoge* (Hebrew: p. 15, 1-18 Davidson; Latin: 13 C); *DSO* VI, pp. 123-24 (Hebrew: pp. 42, 5-43, 8 Hyman; Latin: 11 F-12 A); and I. Madkour, *L'Organon d'Aristote dans le monde arabe* (Paris, 1934), pp. 113-14.
For an almost exact parallel to paragraph 2, see *LDC* II, Comm. 23, 110 G.

7. Averroes makes many references to the validity of syllogisms composed of two affirmative premises in the second figure when the subject and predicate of the major premise are convertible, and to Aristotle's frequent use of such syllogisms. See, e.g., *LP* II, Comm. 23, 57 F; *ibid.*, IV, Comm. 33, 135 D, and Comm. 42, 140 F-H; *IPrA* I, Cap. 1, 2 I; *KMH*, *Heqqêš*, p. 30a, 14-27 (Latin: *De locis Topicis*, 65 H); *QLog* VII, 93 M; *LDAn* II, Comm. 40, p. 193 (Crawford); *ibid.*, Comm. 157, p. 368.

8. *IDC* I, ii, fol. 142a, 20-21 (Latin: 213 B): "... and what can be increased is imperfect"; cf. *LDC* I, Comm. 12, 10 B.

9. On paragraph 3, see *LDC* II, Comm. 23, 110 GH.

10. This would appear to accord neither with Aristotle's definition of "prior in excellence" (*Categories*, 12, 14b, 3-8: "That which is better and more honorable is said to have a natural priority."), nor the version of this definition in *ICat* III, ii (Hebrew: p. 87, 52-54 Davidson [and his note *ad loc.*]; Arabic: p. 112, 6-10 Bouyges / pp. 146, 14-147, 2 Kassem; Latin: 58 E: "The naturally more excellent is thought to be prior to the less excellent...").
Averroes usually states that the more excellent, or the perfect, is prior by nature (cf. *De Caelo* I, 2, 269a, 19-20); or he simply states that it is prior, without qualification (perhaps following *Metaphysics* III, 3, 999a, 13-14, or *De Caelo* II, 4, 286b, 22). See, e.g., *IP* VIII, v, 4, fol. 135b, 11-13:

> Then, since the motion of what is moved in a circle is more perfect than rectilinear motion, and the perfect is by nature prior to the imperfect, what is moved in a circle must be prior to what is moved rectilinearly.

IDC I, iv, fol. 144b, 3-5 (Latin: 274 DE):

> A third demonstration: that circular motion is prior by nature to rectilinear motion because circular motion is perfect, inasmuch as increase and diminution are not found in it; while

> rectilinear motion is imperfect, inasmuch as increase and diminution are found in it.

Cf. *LM* III, Comm. 11, 50 H (Arabic: p. 232, 16-17); *EM* IV (Arabic: p. 131, 3-6; Latin: 386 E; German: p. 109, 5 ff.); *LDC* I, Comm. 12, 9 KL; *ibid.*, II, Comm. 25, 111 G; *LP* VIII, Comm. 58, 399 E-G, and Comm. 59, 399 L.

I believe that the phrase "prior by nature and excellence" is to be explained by reference to *Physics* VIII, 7, 260b, 15 ff., and its treatment by the Greek commentators. Aristotle states there that motion in place is prior to all other motions "in nature, in time, and according to substance (*kat' ousian*)." Themistius (*in Aristotelis Physica Paraphrasis*, *CAG* V,2 [1900; ed. H. Schenkl], p. 225, 18-19), Philoponus (*in Aristotelis Physicorum Libros Quinque Posteriores Commentaria*, [*Excerpta Veneta*] *CAG* XVII [1888; ed. H. Vitelli], VIII, 7, p. 897, 9-11; cf. p. 898, 20-22; and Pines, "Summary," pp. 335-36), and Simplicius (*in Physicorum*, p. 1268, 1-3; cf. similar terminology, *in De Caelo*, p. 41, 13 ff., and the wording "prior in nature and more perfect," *ibid.*, pp. 103, 12 and 107, 27) all interpret "priority *kat' ousian*" to mean "perfection" or "belonging to perfect things," an interpretation probably based on *Physics* VIII, 7, 261a, 13-15 (but cf. *Metaphysics* IX, 8). Moreover, Simplicius (*in Aristotelis Categorias Commentarium*, *CAG* VIII [1907; ed. C. Kalbfleisch], p. 421, 21-24) explicitly identifies "prior in substance (*kat' ousian*)" with "prior in excellence."

Simplicius noted that the meaning of "prior according to substance" in *Physics* VIII is unusual: in *Metaphysics*, "prior according to substance" is synonymous with "prior in nature" (*in Physicorum*, pp. 1269, 10-12, and 1271, 22-25; the reference is to *Metaphysics* V, 11, 1019a, 1-4).

The same interpretation of *Physics* VIII, 7, identifying "prior according to substance" with "prior in excellence," is to be found in the commentary of Abu'l-Faraj, where locomotion is said to be prior in nature, in time, and in excellence (see pp. 884, 18-19; 885, 7-8; 886, 1-2. Abu'l-Faraj evidently depends on Alexander: see Simplicius, *op. cit.*, p. 1268, 3-6). Furthermore, Abu'l-Faraj uses the phrase "prior in excellence" with specific reference to the priority of circular motion. In his commentary on *Physics* VIII, 9, 265a, 17 ff. (p. 912, 12 ff.), he remarks that Aristotle's aim is to prove that circular motion is the first of the motions in place: in time, in nature, and in excellence. Note that the Arabic text of *Physics* VIII, 9, 265a, 22-23, like the Greek, reads: "The perfect is prior to the imperfect in nature, time and *reason* [emphasis mine]," on which Abu'l-Faraj (p. 922, 9-10) writes:

> Because circular motion is more perfect, it is prior in *excellence* to rectilinear motion, for the more perfect is prior to the more imperfect.

In other words, Abu'l-Faraj has identified "prior in reason" with "prior in substance," the latter having previously been identified with "prior in excellence." This is exactly the opinion of Simplicius: "prior in reason" means "prior in substance" (*op. cit.*, p. 1314, 24-25; but cf. Wolfson, *Crescas*, p. 629).

In this *Question*, then, Averroes is following the tradition of earlier commentators. It may have seemed particularly appropriate to him to lay out this view in a treatise dealing with the perfection and priority of the circular figure and circular motion, for Aristotle, at the beginning of a chapter dealing with the perfection and priority of circular and spherical figure (*De Caelo* II, 4, 286b, 10-11), says: "The shape of the heaven is of necessity spherical; for that is the shape most appropriate to its *substance* [emphasis mine] and also by nature primary." (However, Averroes does not follow this interpretation in any of his commentaries on this passage from *De Caelo*, and Themistius [*in Libros Aristotelis De Caelo Paraphrasis*, *CAG* V,4 (1902; ed. S Landauer), p. 66, 14-18] and Simplicius [*in De Caelo*, p. 406, 25 ff.] appear either to have identified "in substance" with the substance of the heavens, or with "in nature.")

In other writings, Averroes exhibits little consistency in his treatment of the terms "prior in reason" and "prior in substance." In *LP* VIII, Comm. 75, 419 D and M, he identifies "prior in reason" with "prior in definition." *Ibid.*, Comm. 56, 397 F and I, he identifies *prius in essentia* with *prius secundum causam*; but *ibid.*, Comm. 58, 399 E-G (to *Physics* VIII, 7, 261a, 13 ff., where Aristotle again speaks of "prior according to substance"), he writes: *et, quia translatio est prior secundum perfectionem aliis motibus....* (See also below, *Question* VII, paragraph 41 where, as in this *Question*, the language of *Physics* VIII, 7, 260b, 15 ff. is applied to the argument of *Physics* VIII, 9, 265a, 13 ff.) Other passages in which Averroes connects "prior according to substance" with perfection are: *LP* V, Comm. 42, 233 EF (to *Physics* V, 4, 228b, 11-15), and *LM* XII, Comm. 44, 327 EF (Arabic: p. 1647, 3), on which see H.A. Wolfson, "The Plurality of Immovable Movers in Aristotle, Averroes and St. Thomas," *Harvard Studies in Classical Philology* LXIII (1958), pp. 244-245; cf. Alexander Aphrodisiensis, *De principiis* (*Fî mabâdî' al-kull*) in *Arisṭû ʻinda-l-ʻArâb* (Cairo, 1947; ed. A. Badawi), p. 267, 18 ff. (Note also *LM* IX, Comm. 13, 240 AB [Arabic: p. 1179, 16-19], where prior *secundum substantiam* is prior *secundum formam*; and cf. *ibid.*, Comm. 15, 241 DE [Arabic: p. 1188, 4-7].)

For a discussion of the meaning of "priority" in Aristotle, see J. Cleary, *Aristotle on the Many Senses of Priority* (Carbondale and Edwardsville, 1988).

11. The best verbal parallel is found not in Averroes, but in Grosseteste, *Commentarius in VIII Libros Physicorum Aristotelis* (Boulder, Colorado, 1963; ed. R. Dales), p. 138: *Perfecta enim imperfecta in eodem genere nature priora sunt....*

Averroes rarely expresses the condition **of the same genus**, especially in connection with the priority of circular motion. See, e.g., *LP* VIII, Comm. 75, 419 E-G, a passage which provides many parallels with our text, where it is said only that in the case of two things, one imperfect, and the other perfect, the perfect is prior to the imperfect. (Cf. *EM* IV [Arabic: pp. 143, 11-144, 9; Latin: 389 C-E; German: pp. 119-120], where the condition is used, but with reference to the relative priority of the separate [incorporeal] intelligences.)

The source for the condition is probably *Metaphysics* X, 4, 1055a, 3-16 (cf. *ibid.*, V, 11, 1018b, 9-10, and V, 16, 1021b, 14-15), which states that that which is greatest in each class [*genei*] is complete [perfect].

(In Averroes, see *LM* X, Comm. 13, 261 G [Arabic: p. 1304, 1-5]; *ibid.*, V, Comm. 16, 120 E [Arabic: p. 570, 9-12]; Averroes interprets "class" as "category.")

Perhaps Averroes has been reminded of the formulation in *Metaphysics* X by the following lines from *De Caelo* II, 4, 286b, 16-17: "But since in any kind [*genei*] the one is naturally prior to the many and the simple to the complex, the circle will be the first of plane figures." Note the treatment of this passage in *LDC* II, Comm. 22, 110 AB [emphasis mine]:

> ... if, therefore, the lines from which the circle is composed were *identical in species* with the lines from which the triangle is composed, then the priority of the circle to the triangle would be like the priority of the simple to the composite. *But actually the circle is prior to the triangle in the same way as the circular line is prior to the straight line: because the circular is perfect while the straight is imperfect... for the circular line is not of the same genus as the straight line.*

(Cf. *LM* XII, Comm. 2, 291 E *et passim* [Arabic: p. 1411, 10 *et passim*]: *Prioritas...in...superficiebus est in eodem genere...*: where the first of the plane figures is not the circle, but the triangle.)

Inasmuch as the specific differentiation of motion is due to the line over which the motion takes place (see below, paragraph 5, and the notes *ad loc.*), the text from *LDC* would imply that circular motion and rectilinear motion do not belong to the same genus. In this *Question*, however, the genus referred to must be locomotion, circular and rectilinear motion being its species (cf. *LDC* I, Comm. 24, 18 M, and *LP* V, Comm. 42, 233 E).

The addition of the condition **of the same genus** in our text may also be connected with the words **by nature, not position** in paragraph 4, below. See below, n. 13.

12. **The syllogism brought to a conclusion by means of an additional proposition**: I have not been able to find this phrase elsewhere in Averroes' writings, except with reference to imperfect syllogisms (syllogisms of the second and third figures; see *IPrA* I, Cap. VI, 15 BC and 16 F; cf. *Prior Analytics* I, 2, 24b, 23-24). The syllogism presented here, however, is in the first figure.

The words of the Hebrew text (*ha-heqqêš ha-môlîd be-tôsepet*) represent the Greek *syllogismos kata proslêpsin*. Proslepsis is usually said to be the minor premise of a hypothetical syllogism (see K. Gyekye, "The Term *Istithnâ'* in Arabic Logic," *JAOS* 92 [1972], pp. 88-92). No hypothetical syllogism is presented in our text.

Alexander (*in Aristotelis Analyticorum Priorum Librum I Commentarium*, *CAG* II,1 (1883; ed. M. Wallies), pp. 263-64), Ammonius (*in Aristotelis Analyticorum Priorum Librum I Commentarium*, *CAG* IV,6 (1899; ed. M. Wallies), p. 68), and Galen (*Institutio Logica*, XIX,1-6 Kalbfleisch) all give examples of Peripatetic usage of proslepsis in the sense of the minor premise of a categorical syllogism. In our text, however, proslepsis applies to the major premise of the syllogism.

As for prosleptic syllogism proper, characterized by a major premise hypothetical implicative syllogism; see C. Lejewski, "On Prosleptic Syllogisms," *Notre Dame Journal of Formal Logic* 2 [1961], pp. 158-176, and J.S. Kieffer, *Galen's Institutio Logica* [Baltimore, 1964], pp. 128-133): it would be possible to phrase Averroes' syllogism in this way, but that does not appear to be what is intended. It is the *addition* represented by the major premise that impresses Averroes.

Professor S. van den Bergh suggested to me (letter, July 9, 1958) that what Averroes refers to here is prosyllogism and episyllogism, the major premise of the episyllogism constituting the additional proposition. This is clearly the only proper interpretation of the passage, but, again, the terminology is a problem. To my knowledge, no Peripatetic writer calls the episyllogism *syllogismos kata proslêpsin*.

It is the commentary of Simplicius to *De Caelo* I, 2 (pp. 38, 14 ff.) which may provide the solution. Here we find a discussion of the ways in which Aristotle's treatment of the priority and perfection of circular motion in *Physics* differs from the treatment found in *De Caelo* (cf., in this *Question*, paragraph 8, below). Simplicius relates the perfection of circular motion to the perfection of the line over which it takes place (in this *Question*, see paragraph 5, below). Then he turns to the question of the priority of circular motion (p. 40, 1-6):

> Positing additionally (*proslabôn*), now, that the more perfect motion is prior, for he says that the perfect is naturally prior to the imperfect; and demonstrating that the circle is naturally prior to the straight line because it is more perfect; and as the lines over which the motion takes place, so are the motions: he has concluded that circular motion is prior to rectilinear motion. Positing in addition (*proslabôn*) to this premise another....

(Cf. *op. cit.*, pp. 236, 7-237, 7, and *Idem*, *in Physicorum*, p. 1313, 27-34.)

The phrase **the syllogism brought to a conclusion by means of an additional proposition,** then, is an interpretation (or misunderstanding) of the use of the term *proslabôn* by Greek commentators to indicate sequence within an argument. In our text, this phrase is equivalent to *proslabôn*, an equivalence elsewhere expressed by the Arabic *zayyad*, a Hebrew term which means "joining" (*ḥibbêr*; see, e.g., paragraph 3 in this *Question*), or a Latin term such as *adiuncta* (e.g., *LDC* I, Comm. 12, 10 D, and *LP* IV, Comm. 42, 140 F-H; note the Arabic translation of *proslambanein* at *Posterior Analytics* I, 12, 78a, 14 [in *Manṭiq 'Arisṭû*, vol. II (Cairo, 1949; ed. A. Badawi)], p. 348, 10], on which see Walzer, "Translations," p. 105). The interpretation of *proslabôn* in our text is most likely that of some translator from the Greek, although it may represent Averroes' attempt to make sense of the text before him. In any case, no Greek Peripatetic could have confused *proslabôn*, as used to indicate sequence, with *syllogismos kata proslêpsin*.

13. The Hebrew term translated as **by position** (*be-hannâḥâ*) represents the Greek *thesei*. I have found no parallel passage in which this term occurs. This makes it difficult to determine which of the many meanings

of *thesei* (see, e.g., Stephanos, *In De Interpretatione Commentarium*, *CAG* XVIII,3 [1885; ed. M. Hayduck], pp. 2, 17- 3, 1) applies here.

Possibly, the term is to be connected with the words **of the same genus** used earlier in this paragraph, and we are meant to think of *Metaphysics* V, 11, 1018b, 9-12:

> The words "prior" and "posterior" are applied (1) to some things (on the assumption that there is a first, i.e. a beginning in each class) because they are nearer some beginning determined either absolutely and by nature, or by reference to something or in some place or by certain people...

It is true that "priority in excellence" is not mentioned in this passage, but some conflation with *Categories* XII would not be particularly unusual. In that case, **by position** would refer to the arbitrary positing of that "first" which determines priority and posteriority (cf., e.g., *EP* IV, p. 57, 15 f. [Arabic]), or to position relative to that "first" (cf. *Physics* IV, 11, 219a, 15 f.), and **by position** would be equivalent to "conventional."

On the other hand, the use of **by position** might be connected with the aim of emphasizing the existential priority of the perfect, rather than its priority in thought. Asclepius interpreted the passage from *Metaphysics* cited above to mean that the prior is first divided into "absolutely prior" and "prior to us and to our *thesin*" (*in Metaphysicorum, Libros A-Z Commentaria*, *CAG* VI,2 [1888; ed. M. Hayduck] p. 323, 17 ff.; cf. *ibid.*, p. 326, 32 where the term *thesei* is used in connection with "prior in knowledge"). For Simplicius (*in Categorias*, p. 421, 14-29), prior *thesei* is divided into what is purely mental (cf. Olympiodorus, *Prolegomena et in Categoriarum Commentaria*, *CAG* XII,1 [1902; ed. A. Busse], p. 147, 14 f.), and into what has some natural priority -- in which case it falls under "prior in order" (*hypo tên taxin*; cf. Stephanos, *loc. cit.*, and Simplicius, *in Physicorum*, p. 1268, 9 f. [Alexander]). According to Simplicius, if "that which is prior in thought" is to have some meaning, it must be the thought of the end (i.e., the completed or perfect thing), and then we have a kind of priority in time: for the thought of the end is prior in time even though its existence is posterior in time:

> But it is first in substance; for the more perfect things are first in substance. This last meaning of 'prior' does not seem to me to be omitted here, as Iamblichos thought, but to be contained in the fourth kind of priority, in which the more honorable and the better are said to be prior.... But he also says the end is prior and better known... even if by nature it is posterior.

Finally, because the subject of this *Question* is the perfection and priority of circular motion, we may refer to Simplicius on *Physics* VIII, 7, 260b, 15 ff. (*in Physicorum*, p. 1269, 5 ff.). Here he points out that Aristotle has discussed only those types of priority useful in discussing motion. As for priority *kata thesin*, if it exists at all in motion, it ought to be taken temporally; or, perhaps, it might be said that the motion belonging to those bodies which are prior *kata thesin* is also prior *thesei*. The latter suggestion seems to imply, again, the identification of *thesis* with *taxis* (cf. *ibid*., p. 1160, 24-27; and Philoponus, *in Physicorum* [*Excerpta Veneta*], p. 897, 10-11).

See also the text cited by F. Rosenthal, "A Commentator of Aristotle," *Islamic Philosophy and the Classical Tradition* (Oxford, 1972), p. 345.

The phrase **by nature or by position** does occur in al-Ghazali (see *TAT* IV, p. 162 [Arabic: p. 273, 15-20; Latin: 71 CD]:

> If you object that souls are not joined to each other, and that they have no order, either by nature or by position, and that you regard only those infinite existents as impossible which have order in space, like bodies which have a spatial order of higher and lower...

This raises the possibility that Averroes might have been thinking of the difference in position between rectilinear and circular motion: The former in the inferior sublunar world, the latter in the superior translunar world.

14. Averroes now proceeds to verify the minor premise in each of the two initial syllogisms. This follows the line laid out by Aristotle at the beginning of *Physics* VIII, 9. Cf. *LP* VIII, Comm. 75, 419 G-M, where we are also told that the major premise is *manifesta*.

15. *Physics* V, 4, 227b, 17-20; *ibid*., VII, 4, 249a, 15-18; *De Caelo* I, 2, 268b, 17-20. See also *LDC* I, Comm. 5, 5 I, and Comm. 12, 9 L; *LP* V, Comm. 33, 227 K; *ibid*., VII, Comm. 30, 330 L-331 A, and Comm. 24, 327 H:

> idest et sequetur quia motus rectus non habet rectitudinem, nisi propter magnitudinem rectam, et similiter motus circularis propter magnitudinem circularem.

EDC I, p. 4, 4-8:

> The simple natural motions are of three types: a motion from the center, and a motion to the center, and these two types are rectilinear motion; and a motion around the center, which is circular motion. Motion differs in this way only because of the difference of the intervals, that is, the circular and the rectilinear....

IDC I, iv, fol. 144b, 6-8 (Latin: 274 E):

> The reason that circular motion is perfect is that the circle is perfect; for it is finite in itself [because] it cannot receive addition or diminution. The reason that rectilinear motion is imperfect is that the straight line is imperfect....

16. See *De Caelo* I, 2, 269a, 19-24; *ibid*., II, 4, 286b, 20-23. *IDC* II, ii, 3, fol. 185b, 14-21 (Latin: 98 FG):

> A second proof, namely: The perfect is necessarily prior to the imperfect, and the perfect is that which is not susceptible of addition or diminution, while the imperfect is susceptible of increase and diminution. The circular line is necessarily perfect because it cannot be added to or diminished and still remain a circular line. But the straight line is imperfect because it can be added to and diminished and still remain a straight line. Inasmuch as the circular line is perfect and the straight line is imperfect, and the perfect is prior to the imperfect, the circular line must be prior to the straight line.

KMH,*Heqqêš*, p. 30a, 16-27 (Latin: [*De locis Topicis*] 65 HI):

> ... and this syllogism is composed in the second figure, because the definition is convertible with itself. For example, we seek to know if the circle is perfect, and we say that the perfect is that to which nothing of its genus can be added, and from which nothing of its genus can be taken away; but the circle is of this description; therefore, the circle is perfect.... For example, we seek to know if the straight line is perfect, and we say that the perfect is that which cannot be increased or diminished; but the straight line can be increased or diminished and still remain a straight line; therefore, the straight line is not perfect. This syllogism is also brought to a conclusion in the first mode of the second figure.

On the perfection of the circular figure or line, see also *EM* I, 362 D (Arabic: p. 29, 6-7; German: p. 23); *EDC* II, p. 43, 14-16; and *LDC* II, Comm. 23, 110 F-H (where mention is again made of the use of two affirmative propositions in the second figure when one proposition is a definition).

17. I find this paragraph very difficult, and I cannot account for the reading of the MSS. I take the **proposition about which doubt occurs** to be the minor premise of the syllogism establishing the perfection of circular motion (see, again, *LP* VIII, Comm. 75, 419 FG, where it is stated that it is the minor premise which requires verification). The verification of this premise is also given in the second figure of the syllogism.

18. Cf. *Posterior Analytics* I, 13; *ibid.*, I, 7, 75a, 28-38.

19. The Hebrew terms used here (*hâ'ôtôt wa-hâ-'êdôt*) represent the Greek *semeia kai tekmêria* (i.e., non-apodeictic arguments). With the exception of his commentaries on the *Organon*, where these terms are explained, this combination of terms is very rare in Averroes. See, however, *EP* VIII, p. 43b, 25 (Arabic: p. 141, 3-4).

In the analysis of arguments, the Greek commentators appear to have blurred the distinction between *semeion*, a sign or indication, and *tekmêrion*, an infallible sign (see Wolfson, *Crescas*, p. 326). For example, in his comments on *Physics* I, 1, Themistius distinguishes only between apodeictic arguments and those "sufficient for us" (*in Physica*,

p. 2, 2-3); in both Philoponus (*in Physicorum*, p. 9, 17 and 27) and Simplicius (*in Physicorum*, pp. 15, 24-25, and 18, 28) the term for the non-apodeictic argument is simply *tekmêriôdês*. This blurring can also be seen in the Arabic translations of Aristotle (see, e.g., the Index of the Arabic version of *De Generatione Animalium* [Leiden, 1971; ed. J. Brugman and H.J. Drossaart Lulofs], s.v. *dalla*, *'alama*, and *šahida*; it is clear that the three terms *semeion*, *tekmêrion*, and *martyrion* are treated as synonyms), and in Averroes (but cf., above, *Question* I, n. 26).

Here, Averroes refers to *demonstratio per signum* (also known as *demonstratio quia*; see *LPsA* I, Comm. 7, 28 B [Mantinus], and *KMH*, *Mopêt*, p. 40b, 23-29 [Latin: *De Demonstratione*, 56 LM], quoted below, *Question* VII, n. 139). The Hebrew terms used for this type of demonstration in this *Question* (*'ôtôt* and *'êdôt*) are unusual. In these *Questions*, the more common term is *mopêt re'âyâ* (see below, *Question* VII, paragraph 38). While there is considerable variation in the Hebrew terms used for *semeion* (e.g., *hôrâ*, *hôrâ'â*, *re'âyâ*, and *sîmân* in *KMH*), as there is in Latin (see, e.g., *LPsA* I, Comm. 7, 28 EF), *'ôtôt* is quite rare (on the term, see Wolfson, *op. cit.*, pp. 446-47), and *'êdôt* is also unusual in the literature. It does occur in the Hebrew text of Maimonides' *Moreh Nebukim* I, 73 (pp. 117b, 14; 124a, 3; 125a, 18), where it translates the Arabic *šâhid*. In a note to 124a, 3, Crescas' comment equates the word with *re'âyâ*. (See also the use of the verbal form *hê'îd* in the Hebrew translation of Themistius, *in De Caelo*, p. 66, 24, and *EDGC*, p. 110, 57 [Kurland].)

Even in the Greek commentators, the use of the double term *semeia kai tekmêria* as the characterization of an argument is not that common. But Simplicius (*op. cit.*, p. 1279, 19-31, to *Physics* VIII, 8, 262a, 6, where Aristotle uses the single term *semeion*) characterizes arguments as *tekmêriôdeis* and *apo semeiou*.

On *demonstratio per signum*, see also below, *Question* VII, n. 139. The fact that, in this type of proof, the middle term is a cause only for our knowledge of the conclusion, and not for the existence of the conclusion (see *LPsA* I, Comm. 8, 30 F), may help to explain Averroes' insistence on the natural priority of the perfect (above, paragraph 4).

20. **Establish**: on the Hebrew term *ha-nôtenîm*, see Wolfson, *op. cit.*, pp. 324-25.

21. I.e., demonstrated apodeictally. If one can put any credence in the Hebrew gender endings, it is the "thing" that is to be so demonstrated, not its form. Presumably, in this case the "thing" would be the unique, eternal and continuous motion.

22. On the phrase **figure and form**, cf. *Categories* VIII, 10a, 11-12. The wording is repeated in the Arabic version of Categories (p. 126, 9 Kassem) and the Hebrew (p. 70, 98 Davidson).

23. We have here a three-fold division of types of proof very different from that found in Averroes' writings on the *Organon* (see, e.g., *KMH*, *Môpêt*, p. 35b, 24-30 [Latin: *De Demonstratione*, 53 FG]), and from what is found in his various commentaries on *Physics* (see, e.g., *EP* I,

p. 9, 7-16 [Arabic]; *IP* I, 2, 434 FH; *LP*, Proeemium, 4 B [E]; *ibid.*, VIII, Comm. 58, 399 G).

24. Lines 38-40 are taken from *De Generatione Animalium* II, 6, 743b, 20-25:

> All the parts are first marked out in their outlines and acquire later on their color and softness or hardness, exactly as if Nature were a painter producing a work of art, for painters, too, first sketch in the animal with lines and only after that put in the colors.

Note that the Arabic version of this text "editorializes" by stating: "... then he adds the colors to those lines so that he paints a perfect animal." This is surely the source for the statement in our text relating color to the perfection of form. Such a statement is not found in *EDGA* II, cap. 4, 85 M-86 A.

It is puzzling to find Averroes using this comparison from *De Generatione Animalium* in a discussion of the various types of proof used in connection with establishing the perfection of circular motion. So far as I know, neither he nor any other commentator ever used this comparison for this purpose. However, Simplicius (*in De Caelo*, p. 56, 2-3 [cf. p. 55, 28 (Xenarchos) and p. 404, 8 ff. (Alexander)]) uses the words: *kai ta poikila de tôn grammôn eidê ouden pros ton logon*. Here, *poikila* means "various," but the word can also mean "color." It seems likly to me that some translator misunderstood Simplicius' (or another, now unknown, commentator's) use of the word and interpreted it in the sense of "painted" or "colored." If Averroes had had before him a text that, in dealing with *De Caelo* I, 2, referred to the disagreement between Xenarchos and Simplicius using the term "colored," it would have been natural for him to have made the connection with the passage from *De Generatione Animalium*.

25. I do not find such a comparison of *Physics* VIII with *De Caelo* elsewhere in Averroes' works. But see Simplicius (*op. cit.*, p. 38, 22-25):

> He also demonstrated in the eighth book of the *Physics* that circular motion is prior to rectilinear motion. But there he demonstrated it because of its perfection and because of its simplicity and absolute continuity, while here the demonstration proceeds from its perfection.

(Note, too, the similarity of the analysis of Aristotle's argument in the lines preceding and following with the analysis found in our text.)

Cf. *IDGC* II, p. 96, 61-62 [Hebrew]: "You ought to know that this discourse functions (*yârûṣ merûṣat*) as the completion [or perfection: *ha-shelêmâ*] of what was demonstrated at the beginning of the eighth book of the *Physics*..." (my translation).

QUESTION III

1. As noted in the Introduction, the text of this *Question* has previously been edited on the basis of MSS. Paris 988, Munich 31, and New York, JTSA 2311 (then Steinschneider 6) by M. Worms in "Die Lehre von der Anfangslosigkeit der Welt." I do not always find his reading of the MSS to have been accurate.

As far as I know, the promised translation of this *Question* by J.L. Teicher ("Avicenna's Place," pp. 46-48; cf. "Averroès inconnu," pp. 337-38) never appeared. An English translation, with notes and commentary, by Barry S. Kogan has appeared ("Eternity and Origination: Averroes' Discourse on the Manner of the World's Existence," in *Islamic Theology and Philosophy* [Albany, 1984; ed. M. Marmura], pp. 203-35.) The text of this treatise presents many problems and offers, therefore, many opportunities for translators to disagree. Professor Kogan's approach to this text is both subtle and ingenious, but it seems to me to miss the point that Averroes is trying to make.

The first fifteen lines of our text are quoted, with some commentary, in Abravanel, *Mip'âlôt 'Elôhîm*, II, 1.

2. Two lists of Averroes' works contain descriptions of treatises which may be identical with this *Question*. The first of these lists (Renan, *Averroès*., p. 464) is to be found in Casiri's catalogue of the Escurial MSS (No. 879) where the description reads:

> A treatise concerning the agreement between the beliefs of the Peripatetics and those of the Mutakallimûn of the scholars of Islam as to how the world exists with respect to eternity and creation.

The second description (Renan, *op. cit*., p. 455), which occurs in Ibn Abi Uṣaybi'a's list, and which corresponds almost exactly with the text of the present *Question*, reads:

> A treatise with reference to the fact that what the Peripatetics believe about how the world exists and what the Mutakallimûn of our religion believe about this is very nearly of the same significance.

Although I believe only the second of these descriptions actually refers to the present work, it is clear from either description that in this *Question* Averroes speaks of coreligionists, not compatriots (*pace* Steinschneider, *HU*, p. 179; *Idem*, "Alfarabi," p. 122; cf. Worms, *op. cit*., p. 66, n. 2). The Hebrew word *mâḥôz*, which may, indeed, have a territorial meaning, is commonly used as a translation of the Arabic *milla* (religion). This may be clearly seen in *Question* VII, paragraph 14, where we have both Hebrew and Arabic text. See also Averroes' commentary on the *Republic*, II, vi, 5 (p. 81 and the note *ad loc.* [Lerner]); but cf. *ibid*., III, xvi, 2 (p. 134 [Lerner]; p. 235 [Rosenthal]), and Goldstein, "Sources," p. 31.

3. The addition to the text is made on the basis of the second of the Arabic descriptions cited in n. 2, above.

4. Cf. *FM*, 11, 15-19 (Hourani). As the wording of this paragraph shows, this treatise is an apologetic work, closely related to *FM* and to *TAT*. It is not unreasonable to conjecture that it may have been written about the same time. On the dating of Averroes' works, see M. Alonso, *Teología de Averroës (estudios y documentos)* (Madrid-Granada, 1947).

5. That is, matter. See *Metaphysics* VII, 7, 1032b, 30-1033a, 1; *ibid.*, VII, 8, 1033a, 24-26.

6. See *Metaphysics* XI, 6, 1062b, 24-33; *ibid.*, XII, 2, 1069b, 18 ff.; and *De Gen. et Corr.* I, 3, 317b, 16-18. Cf. *TAT* I, p. 41 (Arabic: pp. 71, 10-72, 1; Latin: 28 HI); *Kašf*, pp. 50-51 (Cairo, 1935; p.34 [Müller]).

Averroes is not supposed to have known *Metaphysics* XI (Steinschneider, *HU*, pp.162-66). But according to E. Booth, *Aristotelian Aporetic Ontology in Islamic and Christian Thinkers* (Cambridge [England], 1983), pp. 138-39, he certainly knew its general contents, perhaps on the basis of a summary such as that of Nicolaus Damascenus (see H. Drossaart Lulofs, *Nicolaus Damascenus on the Philosophy of Aristotle* [Leiden, 1965]. More likely, in Booth's view, is that his knowledge was based on al-Farabi's "Scopes [*sic*] of the Aristotelian Metaphysics."

7. Cf. *FM*, 11, 20 ff.

8. See *Physics* IV, 12, 221a, 5-13. Cf. *LP* IV, Comm. 118, 192 F; *ibid.*, Comm. 115, 189 M.

9. See *Physics* IV, 12, 221a, 13-29. Note the variants in the Arabic version (p. 449, 7-9) which were apparently unknown to Averroes (see *LP* IV, Comm. 117, 192 HI).

IP IV, iii, 5, fol. 66b, 14-16: "The eternal and perpetual things are not in time, for time does not exceed them, nor does it encompass them as a container."

10. Cf. *TAT* I, p. 31 (Arabic: p. 54, 6-8; Latin: 25 C); *ibid.*, I, p. 50 (Arabic: pp. 85, 11-86, 3; Latin: 31 G); and *LM* XII, Comm. 29, 313 L-314 A (Arabic: pp. 1560, 1-1561, 12).

11. **Prior in nature**: see *Metaphysics* V, 11, 1019a, 1-4 (cf. *Categories* 12, 14a, 30-35), and above, *Question* II, n. 10.

12. See *Physics* IV, 1, 208b, 35-209a, 2; *De Caelo* III, 2, 301b, 32-302a, 9. Cf. *LP* IV, Comm. 7, 124 D.

13. See *Metaphysics* VII, 7, 1032a, 12-14. Cf. *EM* II, 367 AB (Arabic: p. 49, 10-12; German: p. 39); *TAT* III, pp. 89-90 (Arabic: pp. 150-151; Latin: 45 GH).

14. See *TAT* III, p. 127 (Arabic: p. 213, 1-15; Latin: 58 I). This list of characteristics of the "truly created" is, of course, a list compiled by a Peripatetic philosopher. It is difficult to believe that any of the Mutakallimûn would have agreed to this list in every respect, even if these characteristics were not meant to apply to the creation of the world as a whole. In their systems, the atoms would take the place of matter, but the atoms themselves were created *ex nihilo*. The concept of privation has no place in the Kalâm; Averroes can get it into his list only because Arabic has only one word for both the general concept of non-existence and the particular non-existence which is privation. Averroes himself will make use of the fact that the Mutakallimûn certainly do not believe that the world is preceded by time or place, even though they believe the world is created. The only concession Averroes appears to have made to the Kalâm is to speak of an efficient cause which brings what is created from the *possibility* (rather than *potentiality*) of existence to actual existence. "Potentiality," like "privation," implies that the created thing has an inherent nature, or a certain kind of necessity, something which the Kalâm would deny. Cf. the list of characteristics of the created in *FM* 11,15-12,14.

15. Cf. *TAT* I, p. 52 (Arabic: p. 89, 3-7; Latin: 32 BC); *Kašf*, p. 51 (Cairo; p. 34 Müller).

16. **Concomitant to body** (*ḥibbûr le-geshem*): all the MSS read "the connection of body to body" (*ḥibbûr geshem le-geshem*). I have emended the text on the basis of the following passage in *FM*, 12,9 ff.:

> For the theologians admit that time does not precede it [the world], or rather this is a necessary consequence for them since time according to them is something which accompanies motions and bodies.

However, a parallel to the reading of the MSS exists. A. Schmoelders (*Essai sur les écoles philosophiques chez les arabes* [Paris, 1842], p. 163; cf. pp. 164-65) quotes the following definition of time accepted by "les Dogmatiques":

> Le mot *temps* exprime la simultanéité d' un objet avec un autre, ou mieux encore, la connexion d'un certain fait de notre imagination avec un fait connu...

The quotation is presumably from al-Razi's *al-Muḥaṣṣal* (see B. Carra de Vaux, *Gazali* [Paris, 1902], pp. 120-21); unfortunately, I have not had access to the Arabic text of this work.

It is difficult to determine the correct reading of our text without knowledge of the underlying Arabic (and of the Arabic of al-Razi). It would also be useful to know if Averroes here is reproducing the technical language of Ash'arite Kalâm. If he is, it is possible to preserve the reading of the MSS by conjecturing that the term for body was originally *jirm*, the term used for atom. The text would then mean that, for the Mutakallimûn, time is a consequence of the combination of atoms due to God's creative action. This combination is a motion, and time measures it. If, however, the underlying Arabic term is *jism*, and Averroes is using the language of Kalâm, the reading of the MSS would be speaking of the conjunction of two (and not more) atoms, and I do not know what the combination of two such entities would signify. (See the valuable article of R. Frank, "Bodies and Atoms: The Ash'arite Analysis," in *Islamic Theology and Philosophy*," pp. 39-53.)

If we are dealing with the technical terminology of Ash'arite Kalâm, in neither case would the term "body" have the meaning of the celestial sphere. However, the language of our text implies an equivalence between **a consequence of the motion of the sphere** and the expression of the Kalâm which follows. And note the phrase **the concomitance of time with the body of the world** in paragraph 5, below.

17. Cf. *LP* IV, Comm. 43, 141 C.

18. See *De Caelo* II, 1, 283b, 26-29 (on the Arabic version and its sources, see G. Endress, *Die arabischen Übersetzungen von Aristoteles Schrift de Caelo* [Frankfurt a.M., 1966], p. 219); *LDC* II, Comm. 1, 96 A.

IDC II, i, 1, fol. 180b, 16-20 (Latin: 295 DE):

> We say that it has already been demonstrated in one of the preceding chapters that the heavens *in toto* are neither generated nor corrupted, as some men have said; rather, they are perpetual, without a temporal beginning or end; rather, they are the cause of time and they encompass it.

See, however, *Question* V, below, and *TAT* I, p. 37 (Arabic: p. 65, 6-8; Latin: 27 E): "...for it is not of the nature of the Creator to be in time, whereas it belongs to the nature of the world to be so...."

For solutions of this problem, see Wolfson, *Crescas*, pp. 287, 644-46; cf. S. Sambursky and S. Pines, *The Concept of Time in Later Neoplatonism* (Jerusalem, 1971), pp. 63-73, and the notes *ad loc*.

19. Cf. *TAT* III, p. 132 (Arabic: p. 221, 5-9; Latin: 60 E), and *FM*, 11, 17-12, 9.

20. **Finite**: finite in duration. Strictly speaking, the Peripatetic view is that time measures the motion of the world, not its existence.

21. Cf. *TAT* I, p. 32 (Arabic: pp. 54, 14-55, 6; Latin: 25 EF). This problem is also discussed by al-Farabi, "Die Harmonie zwischen Plato und Aristoteles," pp. 36-37 (Arabic: p. 33).

22. There is some doubt as to the correct reading of the text here. Of our five MSS, two read "alteration"; two read "completion"; and one reads "completion of alteration." The reading chosen here reflects the actual position of the Kalâm. The term "completion" has probably crept into the text under the influence of what follows.

23. All the MSS read "which have *no* (*'êyn*) beginning," but from the context, this cannot be correct. Perhaps one should read the relative pronoun *'asher* for *'êyn*).

24. That is, what has occurred in the past has already achieved a definite quantity.

25. I have added the word **absurd** to the text because it seems to be required by the sense. The remainder of this paragraph refers back to paragraph 4. The Kalâm argument as presented here derives from an Aristotelian principle: that the infinite cannot be traversed (cf. *Physics* III, 4, 204a, 3-14; *ibid*., VI, 10, 241b, 10-11; *ibid*., VIII, 9, 265a, 17-20). See Wolfson, *Kalam*, pp. 410-34.

26. This is the terminology of the hypothetical syllogism. Here, however, no formal syllogism is involved. The reference is to the earlier discussion of conditions and priority (paragraphs 3-4).
Averroes is criticizing the Kalâm argument on beginning and completion: that the motions of the sphere are complete and, therefore, must have had a beginning. The refutation of this argument begins at paragraph 6, with the rejection of the notion that the world exists in time at all. Cf. paragraph 8, below, and F. Kholeif, *A Study on Fakhr Al-Dîn al-Râzî and His Controversies in Transoxiana* (Beyrouth, 1966), p. 83, #167 (Arabic: p. 61).

27. On the problem of whether the world as a whole is in place, see Wolfson, *Crescas*, pp. 431-43.

28. All the MSS read: "If the world in its totality does not measure the existence of time...." If this reading were to be followed, we would have a reference to the reciprocal measurement of time by motion (*Physics* IV, 12, 220b, 14-32; cf. *LP* IV, Comm. 133, 205 FG; *TAT* I, p. 49 [Arabic: p. 84, 4-5; Latin: 31 DE/p. 115 Zedler]; Simplicius, *In Categorias*, p. 345, 17 ff. [Iamblichus]; Yaḥya to *Physics* IV, 12 [p. 453, 12 ff.]). Here, however, such a reference seems unsuited to the context, and the emendation required is very minor.

29. See *Physics* IV, 12, 221a, 19-26; cf. *LP* IV, Comm. 117, 192 A.

30. See Yaḥya to *Physics* IV, 12 (p. 453, 16-19).

31. I.e., time is a consequence of the circular motion of the heavens. See *Physics* IV, 14, 223b, 23-33; cf. *LP* IV, Comm. 133, 205 FG; and Simplicius, *op. cit*., p. 345, 27-33 [Iamblichus]. In *Question* II above, Averroes has shown that circular motion is circular because the magnitude over which it takes place is circular.
W. Wieland ("Die Ewigkeit der Welt," in *Die Gegenwart der Griechen im neueren Denkens* [Tübingen, 1960], esp. pp. 300-14), has contrasted John Philoponus' linear view of time with the cyclical view of Simplicius. The Kalâm follows Philoponus, but Averroes stands in the camp of Simplicius. See below, n. 38.

32. *EP* IV, p. 19a, 11-18 (Arabic: p. 63, 8-13):

> It is also possible for a point to be assumed to be a beginning without being an end, or an end without being a beginning, but

> this can only occur in a rectilinear extension in so far as it is divisible and finite and contained. This [assumption], however, is not possible with respect to the instant, for if we take a certain instant, we can only take it to be an end of past time and a beginning of future time. The instant bears more resemblance to a point assumed in a circle than to anything else, for however such a point is assumed to exist in the circle, it is both a beginning and an end.

See *LP* IV, Comm. 125, 198 A-C; *TAT* I, p. 44 (Arabic: pp. 76, 15-77, 5; Latin: 29 LM); Simplicius, *in De Caelo*, p. 43, 12 ff., and p. 44, 15 ff.; *Idem*, *in Physicorum*, p. 1338, 39-40; Abu'l-Faraj to *Physics* VIII, 9 (p. 922, 12 ff.).

33. **Circle**: the reference is to the celestial sphere. The sphere is frequently referred to by this term in the remainder of the *Question*.

34. Cf. *LP* VIII, Comm. 11, 346 KL. At *EM* IV, 385 KL (Arabic: pp. 128, 17-129, 4; German: p. 107), Averroes offers another argument against comparing time with the straight line: that straight lines have position and exist actually, so that they must be finite.

35. Cf. *LP* IV, Comm. 133, 205 DE, and *IDGC* II, iv, 2, 68 (English: pp. 107-08 Kurland; Latin: 388 GK).

36. **Abstract**: that is, without reference to the sphere and its motion. Narboni, *ad Question* V, explains the meaning of **abstract** with reference to time as follows:

> You ought to know that when we say "the number of motion," we mean "every motion," for time measures every motion. And when we say "with reference to the prior and posterior in motion," we mean "in the diurnal motion," for it is the most evident and the most rapid.... It is evident to you from this that time can be taken from two points of view: the first, the true, complete viewpoint is taken from its essence: from time's being a consequence of motion and motion a consequence of a thing moved. This is the true definition of time: the number of motion (in so far as it is a consequence of motion and motion has prior and posterior in it)... and because it is a consequence of motion, time has prior and posterior in it.... The second point of view is the *abstract* form of time, which is the duration of the existence of a thing, without considering whether that duration is the duration of motion in something moved or the duration of the existence of something whose nature it is not to be moved. This is the image of time, not its reality...

Cf. Simplicius, *in Physicorum*, pp. 783, 1-785, 10; Wolfson, *Crescas*, pp. 654-58.

37. See notes 17 and 31, above.

38. Cf. *TAT* II, pp. 70-71 (Arabic: p. 120, 5-14; Latin: 38 IK); *ibid.*, I, p. 39 (Arabic: p. 68, 9-10; Latin: 27 M-28 A).

The argument of the Mutakallimûn is based on Peripatetic principles. It states that present motion may be taken as the end of motion. Since there is an end, there must also be a beginning. If, then, the world is eternal, the motions must be infinite, and an actual infinite will have been traversed. But to traverse an actual infinite is impossible, as is the existence of an actual infinite. Therefore, the world must be created. Although Averroes here attributes the argument to the Mutakallimûn, it is attributed to Plato in *Question* IV, and (in slightly different form) to John Philoponus in *Question* VII. The argument does exist in a work of Philoponus' preserved in Arabic: see Pines, "Summary," pp. 325-36 and 347-52. Averroes may have learned to lump John Philoponus with the Mutakallimûn from al-Farabi. See the Translator's Introduction in S. Pines' translation of Maimonides' *Guide of the Perplexed* (Chicago, 1963), p. lxxxv. So far as I know, it was not al-Farabi's practice to lump Plato with the Mutakallimûn.

See also, below, *Question* IV, n. 5, and R. Sorabji, *Time, Creation and the Continuum* (London, 1983), pp. 211-24.

39. The syllogism would read as follows: If what has occurred in the past had no beginning, it would not have come to an end; but what has occurred in the past has come to an end: therefore, what has occurred in the past had a beginning.

40. See Wolfson, *Crescas*, p. 335.

41. Cf. *TAT* II, p. 72 (Arabic: p. 123, 4-7; Latin: 39 DE); *ibid.*, I, pp. 11-12 (Arabic: pp. 21, 15-22, 11; Latin: 19 C-E). On **end**, see notes 24 and 26, above.

42. **They**: i.e., the Peripatetics.

43. See Yaḥya to *Physics* IV, 12 (pp. 451, 21-452, 3).

44. Cf. Simplicius, *in De Caelo*, p. 43, 13-22 and p. 44, 20-23; *TAT* II, p. 71 (Arabic: p. 121, 3-10; Latin: 38 L-39 A); and *EP* IV, p. 20a, 1-7 (Arabic: p. 67, 9-14):

> For this reason it is said that the eternal things are not in time, for time is not equal [in duration] to their existence, nor does it exceed them at both extremes as it does with those things which exist in it. Furthermore, we have already said that the motion of the uppermost sphere is not in time, for being in time can be predicated of it only in the sense that time is equal [in duration] to its existence but does not exceed it at both extremes of its existence; rather, if we say that the motion of the uppermost sphere is in time, it is in so far as the parts of this motion are in time.

Narboni's commentary to *Moreh Nebukim* I, 74 (p. 20a, 9-30) quotes paragraph 9 of this *Question*, with considerable expansion by Narboni.

As he does in his commentary to this *Question* (quoted below), Narboni goes to great lengths to differentiate between the part of a circle, or of time, when considered as an entity in itself, and the same part when considered as a constituent of a greater whole:

> ...the circle in its totality has no beginning and no end, while each one of its parts is in the circle and has beginning and end: i.e., in so far as they are parts cut from the circle, there then is a point in them which is a beginning and not an end, and a point which is an end and not a beginning. But in so far as that part is mingled in the circle and described by it, what was a beginning and not an end will also be an end, and the point which was an end will also be a beginning, because it is now a point in the circle, not the circle's part, and this circle in its totality has no beginning and no end. Such is the case with motion, and with the totality of created things and the totality of time. For in time, when taken in its totality, any particular instant in it is the end of the past and the beginning of the future.... But when you take a part of time, such as the past complete year... and cut it out [of the circle of time], and give it existence in itself as though no other time will ever follow it, and as though no time ever preceded it, undoubtedly there will be an instant which is an end and not a beginning.... But when you put this year which is part of the circle of time back in time, and put it in whatever place you please in the circle of time, it returns to its nature, and the instant which was an end will also be a beginning.... In general, motion, time, and [the circle of] circular dancers have no beginning and no end. This is because of the circular nature which contains the circle necessitating these things, and which makes them necessary in accordance with its own character, for the All is in a circle...

(On the "circle of circular dancers," see Sambursky and Pines, *op. cit.*, pp. 34-35 and n.1 *ad loc.*; cf. pp. 26-27, and 55.)

45. **Decisively**: lit., "by means of an apodictic proposition."

46. *De Caelo* I, 10, 279b, 12-13: "That the world was generated all are agreed..." See *LP* I, Comm. 60, 36 E, and *LDC* I, Comm. 106, 73 AB.

47. Thus, the world is not created *in* time. Cf. Simplicius, *in De Caelo*, p. 286.

Aristotle asserts (*Physics* VIII, 1, 251b, 15-18; cf. Simplicius, *op. cit.*, p. 296) that of all those who believed the world to be generated, only Plato believed in the generation of time. Cf. *LP* IV, Comm. 88, 174 A; *ibid.*, III, Comm. 31, 99D; *ibid.*, VIII, Comm. 10, 346 E. See also M. Zimara, *Solutiones Contradictionum ...super Octavo Physicorum* (in *Aristotelis... Opera* IV [Venetiis, 1574]), pp. 504-05 [Tertia...].

48. But cf. *TAT* III, p. 100 (Arabic: p. 168, 1-3; Latin: 49 AB). See also, *ibid.*, II, p. 73 (Arabic: p. 124, 8-12; Latin: 39 GH); III,

pp. 96-97 (Arabic: p. 162, 8-12; Latin: 47 LM); III, p. 103 (Arabic: p. 171, 10-11; Latin: 49 M); III, p. 104 (Arabic: p. 172, 15; Latin: 50 C); IV, pp. 156-57 (Arabic: pp. 264, 16-265, 2; Latin: 69 EF). On Averroes' theory of eternal creation, see B. Kogan, *Averroes and the Metaphysics of Causation* (Albany, 1985), ch. 5.

49. The words quoted occur many times in the Quran: e.g., Suras 7, 9, 14, 20. It should not be forgotten that frequently the significant part of a verse (for Averroes' purposes) is the part of the verse that is not quoted. See, e.g., *FM* 13, 4-15.

50. On the necessity for adapting teaching to the different intellectual capacities of men, see *FM*, especially Part III.

51. See *Kašf*, p. 100, 13 ff. (Müller: p. 71); *ibid.*, p. 119, 4-16 (Müller: p. 84); *FM* 9, 7 ff.; *TAT* VIII, pp. 238-39 (Arabic: p. 396, 1-14; Latin: 99 F-H); *LP* I, Comm. 60, 36 C-I.

It is hard to determine if the adjective **scriptural** in our text refers simply to what is said in the Quran, or if it refers to the whole of the religious law. See below, *Question* IV, paragraph 7.

52. This section of the text, beginning with the words **One who does this, however**, occurs at the end of *Question* V in MS. Munich 36 and, according to Steinschneider (*HU*, p. 180; *Hebr. Bibl.* XVII, p. 19), in the Leipzig MS (which was not available to me). I have collated the readings of Mun. 36 with the MSS of *Question* III.

On the general topic of oral teaching of the esoteric, see M. Steinschneider, "Mündlich oder schriftlich?", *Hebr. Bibl.* XVII, pp. 17-19, and for the (alleged) positions of Aristotle and Alexander on this question, see Plutarch, *Lives*, "Alexander" (New York, n.d.; Dryden tr.), pp. 805-06. On the role of oral teaching in the transmission and development of Peripatetic doctrine, see P. Tannery, "Sur la période finale de la philosophie grecque," *Revue philosophique* XLII (1896), esp. pp. 270-72.

Averroes did not always insist on oral teaching as the only permissibile means of imparting knowledge of the esoteric, as we may see from *FM* 17, 12-19 (see also Hourani's n. 142 *ad loc.*, and cf. his n. 191). On the topic of "concealing knowledge," see D. Gutas, *Avicenna and the Aristotelian Tradition* (Leiden, 1988), pp. 225-34, and below, *Question* VIII, paragraph 7.

It is interesting to compare what Averroes says here of the ancient philosophers (on "ancient," see Wolfson, *Crescas*, p. 321) with his remarks on "the first generation [of Islam]" in *FM* 25, 1-5 and 9, 7-12: the latter, who agreed that the Quran had both an exoteric and an esoteric meaning, also agreed that the esoteric meaning ought not to be taught to one who is incapable of learning it; therefore, they did not use their allegorical interpretations in their teaching. Both here and in the passages cited from *FM*, Averroes seems to direct his words against al-Ghazali. He explicitly criticizes al-Ghazali at *FM* 21, 14 ff., where the complaint concerns the overly accessible nature of the latter's books.

The reference in our text to **the exoteric and the esoteric** is not clear. The language is that of Quranic interpretation, but, as applied to the **ancient philosophers**, it may very well refer to the writings of Aristotle.

53. See above, *Question* I, n. 28.

Question IV

1. In his commentary to *Moreh Nebukim* I, 74 (pp. 19a-b), Narboni quotes the first forty-one lines of this text, with some commentary. The quotation is introduced as follows:

> But you must know that Plato greatly magnified this difficulty with respect to the example which these men bring forward. The philosopher Ibn Rushd composed an epistle where he brought forward this objection and showed that Aristotle himself used a similar argument, and [Ibn Rushd] resolved this objection and [also] gave the difference between the proposition which Plato used and that which Aristotle used--for Aristotle accepted it only with conditions. Thus, I have seen fit to cite for you here the truth of what he verified there, and to explain its meaning.

2. See above, *Question* III, n. 47. Presumably, the reference here is to *ordered* motion (cf., e.g., *LM* XII, Comm. 30, 315 FG [Arabic: pp. 1570, 12-1571, 6]; *LDC* III, Comm. 21, 188 EF, and Comm. 24, 191 I) and, specifically, to the motion of the sphere (cf. paragraph 5, below).

3. See above, *Question* III, n. 40.

4. On **completion and end**, see *Question* III, n. 38.

5. This argument is also ascribed to Plato in *EP* (see below, n. 16), but elsewhere Averroes ascribes it to the Mutakallimûn (e.g., *Question* III, above; *Kašf*, p. 53 [Müller: p. 34]; *TAT* I, p. 10 [Arabic: pp. 19, 15-20, 6; Latin: 18 IK]; cf. Judah ha-Levi, *Al-Kitâb al-Khazarî* V, 18, Axiom I [p. 214 in the edition of D. Baneth], and Maimonides, *Moreh Nebukim* I, 71; I, 73, Proposition XI; and I, 74 [the second argument]).
In *Question* VII, the argument (in a sightly different form) is also ascribed to John Philoponus (see *EP* VIII, p. 40b, 10-18 [Arabic: pp. 134, 7-135, 8], where Philoponus is cited as a source for Muslim philosophers; see also *Question* III, n. 38).
For analyses of this argument and its various applications, see R. Sorabji, "Infinity and the Creation," in *Philoponus and the Rejection of Aristotelian Science* (ed. R. Sorabji; London, 1987), pp. 164-78, and Wolfson, *Kalam*, pp. 410-34. On the connection between this argument and Plato's proof for creation, see *Idem*, "Notes on Proofs of the Existence of God in Jewish Philosophy," *Hebrew Union College Annual* I (1924), pp. 585-87.

Averroes offers the following version of this argument in *IP* VIII, iii, 1 (fol. 116a, 19-116b, 4):

> And there is a fourth objection which is well known in our own time: namely, that if there are motions here which are infinite in so far as they have not been without existence in the past, one of two alternatives must be true: Either these [infinite] motions must exist simultaneously, so that there is an infinite [number of motions] in actual existence, which is absurd; or they do not exist simultaneously, but every one of the motions is adherent or successive to another motion preceding it. In this case, however, it follows that the very last motion, at the time we are writing this treatise, for example, cannot exist except after infinite motions. But what has no end has no beginning, and since no first term exists, no final term can exist.

6. *IP* VIII, iii, 2 (fol. 117a, 16-18):

> As for the plural successive motions, an infinite number of them is impossible only in two ways: one, when all of them are actually existent [at the same time]; and two, if the prior motion is assumed to be the cause of the posterior.

Cf. *TAT* I, p. 10 (Arabic: p. 19, 15-20, 6; Latin: 18 IK). On infinite causal chains, see Sorabji, *Time*, pp. 226-31.

7. This is the reading of all the MSS, but the reference is incorrect. Averroes probably had *Physics* V, 2, 225b, 33-226a, 6 in mind (see below), or, possibly, *Physics* VIII, 1 (the MSS [one corrects in margin] also read: "He used the same argument in the *eighth* book of *De Gen. et Corr*..."!) See *LP* VIII, Comm. 15, 350 EF, and *EP* VIII, pp. 40a, 30-40b, 3 (Arabic: pp. 133, 9-134,1):

> It is not possible that there should be a created motion before [every] created motion essentially, for if this were so, no created motion could exist except after the completion of infinite motions. But the completion of what is infinite is impossible, and Aristotle has already demonstrated this in the fifth book of this work...

8. *De Gen. et Corr.* II, 5, 332b, 30-333a, 15. But see also *Metaphysics* II, 2, 994a, 1-7, and *De Caelo* III, 6, 304b, 23-305a, 33. Cf. *TAT* I, p. 11 (Arabic: p. 21, 10-14; Latin: 19 BC). For a discussion of Aristotle's views on the infinite, see J. Hintikka, "Aristotelian Infinity," *Articles on Aristotle* 3 (London, 1979), pp. 125-39.

9. Cf. *TAT* I, p. 56 (Arabic: p. 96, 9-15; Latin: 33 IK); IV, pp. 156-57 (Arabic: pp. 264, 16-265, 2; Latin: 69 EF); II, p. 72 (Arabic: pp, 122, 14-123, 4; Latin: 39 CD); II, pp. 72-73 (Arabic: pp. 123, 15-124, 8; Latin: 39 FG); and *LM* II, Comm. 6, 31 DE (Arabic: I, Comm. 6, pp. 21, 13-22, 1).

10. Cf. *TAT* I, p. 33 (Arabic: pp. 56, 12-57, 10; Latin: 25 KL); IV, pp. 162-63 (Arabic: p. 274, 11-14; Latin: 71 FG); and I, pp. 10-11 (Arabic: pp. 20, 6-21, 2; Latin: 18 K-M).

11. Cf. *TAT* IV, p. 168 (Arabic: p. 284, 5-9; Latin: 73 G).

12. **Supposition**: the Hebrew term is *mâqôm*="place." Here, however, it translates the Arabic *mawḍû‘*, "position, supposition"=*suppositus*, and is an infelicitous choice on the part of the translator. For the Arabic term, see *ICat*, p. 3, 10 Bouyges/p. 71, 10 Kassem, and note the more correct Hebrew term used in the translation (p. 31, 6: *ha-mûnâḥîm*).

13. Cf. *TAT* IV, p. 158 (Arabic: pp. 267, 12-268, 3; Latin: 70 AB); *LP* V, Comm. 13, 218 H-K; *ibid.*, VIII, Comm. 15, 350 C-E; *Kašf*, pp. 53-54 (Müller: pp. 34-35); and *LM* II, Comm. 6, 31 DE (Arabic: I, Comm. 6, pp. 21, 13-22, 2).

14. The heat produced by the celestial motion is discussed by Aristotle at *De Caelo* II, 7, 289a, 16-23, and *Meteorologica* I, 3, 340b, 12-14 and 341a, 19-30. To my knowledge, Averroes never refers to the heat produced by the motion of the heavens in any other passage where his intent is to establish the possibility of infinite, successive motions.

In this text, does Averroes have in mind that this heat is responsible for the motion of the elements (see *EM* IV, 393 H [Arabic: p. 162, 9 ff.; German: p. 136]), and, therefore, even a "last" motion of the heavens would not be the end of motion? I do not find such an interpretation in Averroes' commentaries to *De Caelo* II, 7 or *Meteor.* I, 3 (but cf. *LDC* II, Comm. 42, 126 KL), nor have I found it in the commentaries of Themistius and Simplicius to *De Caelo*.

15. *TAT* I, p. 10 (Arabic: p. 19, 9-12; Latin: 18 HI)

16. *EP* VIII, p. 40b, 6-9 (Arabic: p. 134, 4-6):

> Plato, and those who followed him of the Mutakallimûn of the men of our time, and all those who asserted the creation of the world, thought that what is accidental was essential. Thus they denied that there could be motion before motion *ad infinitum*. [The Arabic text reads: Plato, and those who followed him of the Mutakallimûn of our religion and of the Christian religion, and...]

Cf. *LP* VIII, Comm. 47, 388 KL, and *LDC* I, Comm. 120, 81 M-82 B (where, however, the subject under discussion is not motion).

17. On the term **nature** (*ḥôq*), see Wolfson, *Crescas*, p. 388, n. 146.

18. Cf. *LP* VIII, Comm. 4, 341 HI, and Comm. 15, 349 H-350 A. See also, below, *Question* V, paragraph 5; *Question* IX, paragraphs 3 and 7; Sorabji, *Time*, pp. 238-245; and H. Hugonnard Roche,"L'Épitomé du *De*

Caelo d'Aristote par Averroès: questions de méthode et de doctrine," *AHDLMA* 51 (1984), especially pp. 31-35.

19. **Argument**: The Hebrew term (*re'âyâ*) implies a non-demonstrative proof. See above, *Question* II, n. 19, and below, *Question* VII, n. 139.

20. The argument here must be read in the context of the previous paragraph: even if Plato's argument is accepted, and one agrees that motion in this world must have a beginning, it is still possible to preserve the immutability of the mover and the eternity of motion by positing the existence of an infinite series of worlds prior to this world; the notion of an eternal action of creating is more acceptable than the notion of a single creation.

On the theory of the infinite succession of worlds, see *De Caelo* I, 10, 280a, 23-28, and n. 25, below.

21. Cf. *LP* VIII, Comm. 6, 342 CD; *TAT* I, pp. 17-18 (Arabic: pp. 32, 4-33, 6; Latin: 21 G-K / pp. 87-88 Zedler). For the connection of the argument from possibility with the theory of an infinite succession of worlds, see *TAT* I, p. 58 (Arabic: pp. 98, 14-100, 1; Latin: 34 D-G).

The argument here depends on the proposition that possibility requires a substratum. The possibility of this world, if it is created, would then have to exist in some other thing. If it is assumed that that other thing is also created, and so on, an infinite regress results. In this case, the assumption of an infinite series of worlds does not solve the difficulty: an infinite series of possibilities cannot result in an actuality.

Possibility in the agent is alluded to in paragraph 6 above, and paragraph 8, below. Cf. *TAT* I, pp. 59-60 (Arabic: pp. 100-102; Latin: 34 H-K).

22. I.e., in the existence of infinite, successive worlds.

23. See above, *Question* III, n. 51.

24. **Literal**: The Hebrew *nigleh* represents the Arabic *ẓâhir*. Both terms also carry the sense of "exoteric" (see above, *Question* III, paragraph 12).

25. The translation **tradition** depends on vocalizing the Hebrew term as sippûr. Should one vocalize the word as sâpûr, the translation would be, "the various books and compilations [of traditions] have already *counted* a world both before and after...". In either case, a *finite* sucession of worlds is implied, and this was actually the opinion of some Muslim thinkers. See H. Wolfson, "The Platonic, Aristotelian, and Stoic Theories of Creation in Hallevi and Maimonides," *Essays in Honour of the Very Rev. Dr. J.H. Hertz* (London, 1942), pp. 432-36 (especially, p. 434, for the Muslim view). See also Sorabji, *Time*, pp. 249-52.

Narboni, *ad loc.*, points out that Averroes' words apply to all the religions of his time, and he goes on to quote a Jewish tradition (*Bereshit Rabbah* III, 7) that before the creation of this world, God was building worlds and destroying them. For Narboni, the significance of the tradition is that it reveals the nature of motion and time, both of which require something in motion.

Averroes (*IP* VIII, i, 2, fol. 111a, 8-12), following Aristotle (*Physics* VIII, 1, 250b, 18-23), also sees one's opinion on the number of worlds as significant of one's position on the nature of motion:

> For this reason, their opinions concerning the perpetuity of motion, or its generation and corruption, agree with, and are consequences of their opinions concerning the world. For one who believes in infinite worlds which are generated and corrupted, believes in the perpetuity of motion as a genus [but not in the perpetuity of the motions of this particular world]; while one who believes that the world is one, and that it is generated and corrupted, accordingly believes that motion has the same type of existence.

Cf. *LP* VIII, Comm. 1, 338 L.

26. Cf. *Kašf*, p. 74, 1-5 (Müller: p. 49).

27. The text at this point is corrupt, but the meaning of the sentence is clear.

28. See *TAT* [About the Natural Sciences I], p. 332 (Arabic: p. 541, 5 ff.; Latin: 135 D), and van den Bergh's note *ad loc.*; *Kašf*, pp. 120 f. (Cairo).

29. Cf. *TAT* I, p. 3 (Arabic: p. 8, 7-12; Latin: 16 BC), and Sorabji, *op. cit.*, pp. 253-267.

30. **Explained**: the Hebrew root is *p.r.d*, rather than the expected *p.r.š*. Both roots, however, translate *f.ṣ.l*, with all the nuances of that Arabic root. See below, *Question* VI, paragraph 3; *Question* IX, paragraph 9; and cf. Endress, *Arab. Übersetzungen*, p. 260 (to *De Caelo* III, 4, 302b, 20: where the Greek reads *deiknysthai*, and Endress conjectures that the prototype of the Arabic version read *diaireisthai* [cf. *ibid.*, p. 198, to *De Caelo* I, 11, 280b, 2-3]).

Question V

1. Steinschneider conjectured (*HU*, p. 180; "Alfarabi," p. 112) that this *Question* might be identical with the treatise (*mas'ala*: lit., "question") on time listed by Ibn Abi Uṣaybiʿa (Renan, *Averroès*, p. 456; cf. *ibid.*, p. 458). It should also be noted that the Escurial list, which does not contain a reference to a treatise on time, does refer to a treatise (*maqâla*) on the eternal (*sarmadî*) existence and the temporal existence (*ibid.*, p. 464).

Steinschneider also noted (e.g., *HSS-Verzeichniss...Berlin*, p. 90; *Hebr. HSS...in Muenchen* [2nd ed.], p. 22) that in the margin of his own

MS, this *Question* was designated as *ha-ma'amâr ha-kôlêl*. If, indeed, this Hebrew phrase translates the Arabic *al-maqâla al-jamî'* ("the inclusive treatise"; see G. Vajda, *Recherches sur la philosophie et la kabbale dans la pensée juive du Moyen Âge* [Paris, 1962), p. 120), this designation would appear more properly to belong to *Question* III, or perhaps *Question* VII. It is possible that the annotator was influenced by a phrase in Narboni's commentary:

> When he says that **time is a term referring to the duration of existence of things subject to motion**, he means by this "true time" which in its nature *includes* both time's matter and its form.

Narboni himself, however, does not use the term "inclusive treatise" for this work; he reserves it for *Question* VII.

2. Even though all our MSS use the term *halîkâ* for **duration**, the term found in Narboni's commentary *ad loc.*, and in his quotation from paragraph 2 in his commentary to *Moreh Nebukim* II, Introduction, Proposition XV, is *hemšêk*. Both Hebrew terms translate the Arabic *imtadâd* (or *mumtada*) which, in turn, translates the Greek *diastêma*. For a translation of Narboni's quotation and his explanation of it, see H. Wolfson, "Note on Crescas' Definition of Time," *Jewish Quarterly Review*, N.S. X,1 (1919), pp. 13-14. On time as "duration" or "extension" in late Greek writers, see Sambursky and Pines, *Time*, pp. 33, 37, 91; and R. Sorabji, *Time, Creation and the Continuum*, pp. 81-83.

3. On the evolution of definitions of time, see Wolfson, *Crescas*, pp. 638-40; *Idem*, *The Philosophy of Spinoza* (Cambridge, Mass.; 1948), I, pp. 331-46. The description of time as "duration" (or "extension") by Averroes is, so far as I know, peculiar to this *Question* and the following passages from *TAT*: I, pp. 49-50 (Arabic: pp. 84,4-85, 1; Latin: 31 DE); I, p. 50 (Arabic: p. 86, 3-8; Latin: 31 GH); I, p. 52 (Arabic: p. 89, 9-13; Latin: 32 CD); II, p. 70 (Arabic: pp. 119, 15-120, 3; Latin: 38 HI / p. 137 Zedler). Cf. Ibn Ḥazm's definition of time (M. Asín Palacios, *Abenhazam de Córdoba* (Madrid, 1927-32), v. 2, p. 104).

4. *EP* IV, p. 17a, 15-17 (Arabic: p. 59, 10-11; cf. p. 58, 10-16):

> ... for we are not able to conceive of it [time] apart from motion, while it is possible for motion to be conceived of apart from time.

IP IV, iii, 4, (fol. 62a, 20-23):

> ... and inasmuch as it is evident from the nature of time that it does not exist apart from motion and change--for that which is not comprehended in motion is not comprehended in time--but when we have comprehended time we have already comprehended motion...

Cf. *LP* IV, Comm. 97, 178 A; *TAT* I, p. 37 (Arabic: p. 65, 4-5; Latin:

27 DE); *ibid.*, p. 42 (Arabic: p. 74, 2-6; Latin: 29 DE); Alexander Aphrodisiensis, *Treatise on Time* (in A. Badawi, *Commentaires sur Aristote perdus in Grec et autres épitres* [Beyrouth, 1971], p. 23, 1-4).

5. See van den Bergh's n. 3 to p. 166 of his translation of *TAT* (v. 2, p. 105).

6. See Wolfson, *Spinoza* I, pp. 358-69; Sorabji, *op. cit.*, pp. 98-136.
In spite of the emphasis on the specific meaning of the term *dahr* in this *Question*, it should be noted that Averroes himself makes use of a large number of terms for "eternity," and that he applies them with no special consistency (all references to *TAT* and *LM* are to the Arabic text): e.g., the use of *sarmadî* in *LM* XII, Comm. 10, p. 1447 and *ibid.*, Comm. 29, p. 1557, where it refers to what is not subject to motion, whereas in *LM* XII, Comm. 29, pp. 1558-59 the term refers to what is moved, but moved eternally. (Cf. *ibid.*, Comm. 41, p. 1633; Comm. 30, p. 1565.) *Dahr*, too, does not necessarily have the technical meaning ascribed to it here or in *TAT* II, p. 120. Cf. *LM* III, Comm. 15, p. 258; *TAT* I, p. 95; *IDGC*, p. 92, 86; *EDGC*, p. 122, 75. Cf. also Aristotle's discussion of *aiôn* at *De Caelo* I, 9, 279a, 18-28.

7. See *Physics* IV, 12, 221b, 15-23. Cf. *LP* IV, Comm. 118, 193 BC; *ibid.*, Comm. 119, 193 F; *EP* IV, p. 67, 7-11 (Arabic); and the citations from *IP* in Wolfson, *Crescas*, pp. 645-46.
Narboni *ad loc.* comments that the term "eternity" refers to the second aspect of time--not true time, but its likeness, and that the "things not subject to motion" are the separate [incorporeal] beings which exist perpetually and whose existence has duration, but not a duration which can be called time.
Note that Narboni's definition of eternity would imply that the sphere and, therefore, its motion are in time. This, however, is not the position of Averroes, as is clear from *Question* III, and other passages cited in the notes to that *Question*. Cf. *TAT* I, p. 39 (Arabic: p. 65, 7-10; Latin: 27 E).

8. The Hebrew preposition (*be*) which I have translated as **in** is ambiguous: its meaning can be either locative or instrumental. My translation is based (1) on Averroes' terminology in *TAT*, where the instrumental preposition is usually the Arabic *'an*, and this usually following a verb for being generated or emanated (cf., in the Arabic text, XIII, p. 466, 16 ff.; XII, p. 452, 12 f.; IX, p. 406, 5-9). No such verb occurs in our text. Given the very literal translation of these *Questions* (as may be seen from *Questions* VI and VII below), it would be extraordinary for such a verb to have been omitted. (2) Again, in *TAT* XIII (p. 467, 9-10 [Arabic]) we have reference to the impossibility of temporal beings existing in (*fî*) an eternal being. (3) Narboni, in his commentary, seems to understand the preposition in a locative sense:

> For Averroes further proved that the incorporeal things cannot be moved, just as those things which have their existence in

> motion cannot become eternal, i.e. unmoved. And he said, **If the things which are in the eternal existence cannot be in the existence subject to motion,**--for they are not bodies that can be moved --**then the moveable things cannot exist in the eternal existence,**--i.e. so that they would be consequences of the perfectly eternal existence, without motion; for motion belongs to their nature and essence, and motion is the opposite of eternity. For the essence of motion is either the generation of the thing itself or the generation of one of its antecedents. But generation and actualization are opposed to eternity.... Thus, the two existences are separate.

9. This line of argument is analogous to that used by Aristotle in *De Caelo* I, 12, and by Averroes in *IDC* I, x, 2, 6 (fol. 173b, 9-14; Latin: 290 GH):

> After it had been demonstrated to him that the eternal is neither generated nor corrupted; and that what is generated and corrupted is not eternal; and that the two are natures separate one from the other in the maximum degree: he began also to demonstrate here that everything that is not generated is eternal and, similarly, everything that is not corrupted is eternal. This is the contrary of what was first demonstrated: that the eternal is neither generated nor corrupted.

In this *Question*, **separate** means that the existences are distinct from one another in kind, not degree. Cf. *LM* I, Comm. 39, 23 A (Arabic: p. 138, 19-139, 2).

10. That is, the sphere, even though subject to motion, has no beginning for its existence or its motion.

Clearly, there is something wrong with the text at this point. One may derive smoother readings from the MSS, but this will not solve the problem: Averroes has not yet given grounds for concluding that both existences must be eternal. That proof begins in paragraph 4, introduced by a phrase which implies that another proof has been given earlier.

Narboni expatiates on the meaning of the term "eternity" as applied to things capable of motion, but he is no help at all on the construction of the argument.

11. Although only one MS contains the word **void** in the text of Averroes, all those MSS containing the commentary of Narboni have this reading in the quotation of the text in the commentary:

> **And even if the existence capable of motion is eternal, it is not possible that it should be void of that disposition,** i.e. motion, for motion is like its form and its essence, in so far as it is a being capable of motion. This is why he says, **since nothing,** etc.

Cf. *TAT* III, p. 103 (Arabic: p. 172, 3 f.; Latin: 50 A); *ibid.*, IV, p. 156 (Arabic: p. 264, 11 f.; Latin: 69 D); *LP* VIII, Comm. 1, 338 HI (cf. 339 D); *DSO* IV, p. 117 (Latin:10 I; Hebrew: p. 49, 32-36); *ibid.*, VII, 14 BC; *Question* IX below, paragraph 6.

12. Narboni, *ad loc.*:

> **Because there are only these two existences** means that if motion had not been created in the existence capable of motion, and something in motion actually existed, it would have had to have been created in the eternal existence.

13. Cf. *TAT* I, pp. 42-43 (Arabic: p. 74, 6-10; Latin: 29 E), and, for the reverse, *LDAn* III, Comm. 5 (p. 391, 140-43 Crawford). Cf. also *Question* IV, paragraph 6 and n. 17. The point is that the eternal contains no possibility of any kind. Note that Averroes does not even mention the alternative of creation *ex nihilo*.

14. Narboni, *ad loc.*: "**Conjoined to the eternal existence**, means the relation of that which is moved to the mover."

According to *TAT* VIII, p. 237 (Arabic: p. 394, 4-9; Latin: 99 BC), the link between eternal existence and temporal existence is the movement of the sphere. Cf. *LDC* I, Comm. 22, 17 LM; *LP* VIII, Comm. 15, 352 B; *TAT* XIII, p. 284 (Arabic: p. 469, 2-9; Latin: 115 M-116 A). For the wording of our text, cf. Judah ha-Levi, *Khazarî* I, 1 (p. 4, ll. 5-7 Baneth).

Note that Averroes does not specify the type of causation involved, nor the type of priority that would be entailed.

15. The reference to metaphysics is puzzling. Averroes was critical of Avicenna's purely metaphysical proof for the existence of the Prime Mover for the very reason that it was metaphysical. (Perhaps the use of metaphysics here is legitimate because the existence of the Mover is not at issue, but only his relation to the world.) See, e.g., *LP* I, Comm. 83, 47 FG; *ibid.*, VIII, Comm. 3, 340 EF; *LM* XII, Comm. 5, 293 B-D (Arabic: pp. 1422, 14-1424, 4); *TAT* VIII, pp. 237-38 (Arabic: pp. 394, 6-396, 1; Latin: 99 B-F); H. Wolfson, "Averroes' Lost Treatise on the Prime Mover," *Hebrew Union College Annual* XIII,1 (1950-51), pp. 687-97.

In *TAT* I, pp. 38-39 (Arabic: pp. 66, 6-68, 2; Latin: 27 H-L), which provides an interesting parallel to our text, the proofs cited are all physical. See, however, *LM* XII, Comm. 29, 313 E-I (Arabic: pp. 1557, 6-1559, 7); cf. *Metaphysics* XII, 1, 1069a, 30 ff.; *ibid.*, XII, 6, 1071b, 3-5.

16. Cf. above, *Question* III, paragraph 2.

17. Narboni comments that this is because the existence capable of motion exists only in motion, and motion comes from a mover. Thus, that which gives motion is that which gives existence. See also nn. 8 and 12, above.

18. Cf. *TAT* II, p. 73 (Arabic: p. 124, 11-12; Latin: 39 G).

19. Cf. B. Lewin, "La notion de *muḥdath* dans le Kalam et dans la

philosophie," *Orientalia Suecana* 3 (1954), pp. 91-93 (Ibn Suwar).

20. Cf. Alexander, *De Principiis* (p. 262 Badawi):

> The mover of the divine body is prior to all existing things: it is prior to the generated and corrupted things in time, but it is prior to the divine body in rank, not in time, as is the case with the things that exist simultaneously, i.e., the cause and the effect.

21. See above, n. 13, and *Question* IX, below. On the modal terms in Aristotle, see J. Hintikka, "Necessity, Universality, and Time in Aristotle," *Articles on Aristotle* 3, especially pp. 110-115.

22. Quran 11:9 (Rodwell). According to *TAT* III, p. 132 (Arabic: p. 222, 5; Latin: 60 F [but lacking the Quranic verse]), the literal meaning of the verse is that the heavens are the first of created things. Cf. Worms, *Lehre*., pp. 30-31.
Narboni connects this verse with Gen. 1:2:

> **And His throne had stood ere this upon the water**, is like the saying of the Torah, "And the spirit of God hovered over the face of the waters." I.e., the connection of that which hovers and that hovered over.

23. Unfortunately, the commentary of Narboni does not take up this puzzling paragraph--perhaps an indication that it does not belong in this text. The material presented here is reminiscent of *De Caelo* I, 12 (cf. *Metaphysics* IX, 8) where the argument applies to the eternity of the sphere. The majority of our MSS, however, specify "eternity" by the term *dahr*, a term which cannot be applied to the sphere, according to what was said earlier in the text. Even if we omit the term *dahr*, as one MS does, it seems at least questionable that Averroes would describe the sphere as existing *in* eternity.
The term *dahr*, of course, can be applied to the Prime Mover and the intelligences. However, I can think of no good reason why Averroes should apply this argument to them.

24. Instead of **truly** (*be-'emet*), three MSS read *b'nwt*. In different writings ("Mündlich," *Hebr. Bibl*. XVII [1877], p. 19; *HU*, p. 180), Steinschneider chose different readings. It might be better to read with the one MS that omits the word altogether, but the sentence would still be difficult grammatically and in meaning. One would expect to find here a sentence to the effect that the reason something cannot be generated and corrupted at one and the same time is because generation and corruption are contraries, and contraries cannot exist in one and the same subject at one and the same time.

25. It is at this point in the text that MS. Munich 36 contains the lines which other MSS include in the text of *Question* III. See *Question* III, n. 52. It should be noted that MS. Munich 36 does not contain the text of *Question* III.

26. Here, again, the honorific attached to the mention of God is abbreviated in the Hebrew text. It could be almost anything, ranging from "blessed be He!" (the common Jewish expression) to a translation of the common Muslim expression "may He be exalted!"

It is interesting to note that in *Question* III, where he dealt with the equivocal nature of the term "created," Averroes ended the treatise with a reference to God directing us to the truth. In this treatise, dealing with the equivocal nature of the term "eternal," he closes with the words, "God knows the truth."

QUESTION VI

1. This work is most probably identical with the treatise on seed and sperm (*maqâla fî al-buzûr wa-al-zurû'*) mentioned in the Escurial List (see Renan, *Averroès*, p. 464; Steinschneider, *HU*, p. 180). The description of the unique Arabic MS of this *Question* given by Derenbourg in his catalogue of the Escurial MSS (v. I, p. 440) is erroneous, clearly due to the difficulty of deciphering this MS. Steinschneider (*loc. cit.*) errs in transcribing the end of the *Question*, reading *be-dibbûr* for *be-kâbôd*. This *Question* is not to be confused with the treatise *De Spermate* found in the 1560 Latin edition of Averroes' works (*HU*, p. 180, n. 545; Renan, *op. cit.*, p. 78).

On the nature of Aristotle's discussion of animal generation, see R. Sorabji, *Necessity, Cause, and Blame* (Ithaca, N.Y., 1980), pp. 166-174.

2. Although material pertinent to the topic of this *Question* is treated many times by Averroes, this is the only work I know of where Averroes discusses the powers in semen without reference to related problems such as the role of the Active Intellect, or the Platonic forms, in generation; spontaneous generation; the role of the female, etc. The primary Aristotelian text on which the *Question* is based is *De Generatione Animalium* I, 18-II, 4 (Narboni's commentary refers frequently to "chapter 16"; for the numbering of chapters and books of Aristotle's works on animals in the Arabic and Hebrew versions, see the Arabic text of *De Generatione Animalium* [*Fî Kawn al-Ḥayawân*; Leiden, 1971; ed. J. Brugman and H. Drossaart Lulofs], Introduction, p. 1). Averroes, however, does not mention his source under any title here.

Because of the narrow focus of this *Question*, Averroes ignores certain topics treated by Aristotle in those chapters, topics which are essential to the development of Aristotle's argument. For this reason, Averroes does not always appear to agree with his source. On the other hand, *EDGA* omits material discussed in this *Question*. (Note that Narboni and Gersonides refer consistently to Averroes' *bî'ûr* on *De Gen. An.*, and that the Latin title of the commentary is *Paraphrasis*. Both the Hebrew and Latin terms usually refer to a Middle Commentary. However, it is now generally agreed that this work is the Epitome [*HU*, p. 144, n. 258; F. Peters, *Aristoteles Arabus* (Leiden, 1968), p. 45].)

As Gersonides noted (*Milḥâmôt ha-Šem* V, 3, i), the significant sources for Averroes' views on generation are *EDAn*, *EDGA*, *LM* VII and *LM* XII.

All these sources provide parallels to the text of this *Question*. *Colliget* II, 10, 22 H-23 L might be added to the list.

Narboni, at the beginning of his commentary to our text, explains that he will comment on this treatise (which he later calls "the precious treatise on semen") at greater length than usual because (a) it investigates the creation of animals, especially man, and it is the natural order that man should know his own origin and then include other knowledge; and (b) this treatise includes what was proved in *EDGA* in greater depth.

3. One Hebrew term (*zera'*) is used here in a double sense to cover two Arabic terms (*buzûr*/seed; *zurû'*/semen). The double sense of this Hebrew term is not uncommon; see the index to the Hebrew texts of *IDGC* and *EDGC* s.v.

4. I take this to mean that it is the semen and the seed which resemble that which possesses semen. See below, paragraph 9.

5. See *De Gen. An.* II, 1, 734a, 3-5, and cf. *LP* III, Comm. 17, 92 CD.

6. Cf. *LM* VIII, Comm. 12, 220 BC (Arabic: p. 1075, 12-14):

> ... et causam moventem quae est sperma maris: et causam, quae est secundum formam (*ka-al-ṣûra*), et quae dat quiditatem (*mâhiyya*) rei per quam est...

Cf. also Ibn al-Samḥ to *Physics* II, 1 (p. 81, 13-14):

> ... here he means by "nature" the form of the thing through which the thing is what it is, and this form determines its substance, like the form of the semen.

7. I.e., specifically, but not numerically, identical. See *De Anima* II, 4, 415b, 3-7.

Of the species: These words are found only in the Arabic MS. This reading is supported by the Greek text of *Metaphysics* VII, 7, 1032, 22 ff., but not, apparently, by the Arabic version. See *LM* VII, Text. 22, and Averroes *ad loc.* (Arabic: p. 841, 3-6; Latin: 172 H); *EM* II (Arabic: p. 49, 11-18; Latin: 367 B; German: pp. 39-40). Cf. *Metaphysics* VIII, 8, 1049b, 18 ff., where, again, the Arabic differs. Averroes' source for these words may have been *Physics* II, 7, 198a, 24-27.

8. See *De Gen. An.* I, 20, 729a, 9-11.

9. Both the Arabic and Hebrew texts use terms which normally mean "first." Here, however, both terms represent the Greek *euthys*. Cf. *De Gen. An.* I, 21, 729b, 3 f., and n. 13, below.

That some body produced by the father generates the embryo would imply a material contribution by the father, a notion rejected in *De Gen. An.* I, 22 (cf. *ibid.*, II, 3). Note, too, that in paragraph 9, the semen, or its power, is described as **instrument**. Elsewhere, we are told that an instrument is not part of the thing made (*EDGA* II, cap. 3, 76 H).

10. This is the reading of the Arabic, followed by Narboni in his commentary, and by MS. Munich 31. The remaining Hebrew MSS read, "so that the embryo exists in the womb."

11. Cf. *EM* II, 367 F (Arabic: p. 50, 18-19; the German translation does not contain this passage).

12. **Which separates itself from the father**: This is the reading of the Arabic. The Hebrew MSS (and Narboni) read the opposite: that the power is not separated from the father. This discrepancy is most likely due to the orthographic similarity of the Arabic terms for "separated" (*infaṣala*) and "attached" or "continuous with" (*ittaṣala*), plus the frequent occurrence of the latter term in this part of the text. It is possible, however, that all the MSS need correction, and that we should read: **a power from the father which attaches itself to the semen**. Cf. *LM* V, Comm. 5, (Arabic: pp. 508, 12-16, and 509, 7-12; Latin: 107 I-L).

13. **The proximate mover**: The Hebrew and Arabic terms (*ha-mênîa‘ ha-ri’shôn/al-muḥarrik al-’awwal*) more commonly refer to the prime mover, but not here, nor in paragraph 9, below. Cf. *De Gen. An.* II, 1, 732a, 2 f.; *ibid.*, IV, 1, 765b, 13 f.; *Physics* II, 7, 198a, 26; and *TAT* III, p. 126 (Arabic: p. 211, 9-10; Latin: 58 E), where generating man is explicitly said to be the proximate agent (*al-fâ‘il al-qarîb*) of generated man.

According to Averroes, the prime mover in generation is the sun and/or the Active Intellect (see *LP* V, Comm. 13, 218 K); or the celestial body (or bodies; see *ibid.*, VIII, Comm. 15, 352 B; *ibid.*, Comm. 46, 387 M-388 B; *ibid.*, Comm. 47, 388 HI:

> Et modus generationis istorum, quae movent se adinvicem, per accidens est. quia sunt quasi instrumenta, per quae movent corpora coelestia, s. quod homo non dat in generatione hominis, nisi illud, quod est quasi instrumens: et corpora coelestia sunt quasi artifex, et virtutes informativae... fiunt etiam à corporibus coelestibus.

But see *LM* V, Comm. 1, 101 B (Arabic: p. 478, 10-16), where the parents are said to be remote, extrinsic causes; and *EM* II, 367 CD (Arabic: p. 50, 6-14; German: p. 40).

14. See *De Gen. An.* I, 21, 729a, 34-729b, 8; cf. *ibid.*, II, 1, 733b, 31-734a, 1.

15. See *EDAn* I, p. 6, 7-8; *EDC* II, p. 41, 10; Gersonides, *op. cit.*, VI, i, 3, p. 299: " We have not found this argument in Aristotle explicitly, but we have derived it from the implications of his discourse." Cf. Philoponus, *in Physicorum*, pp. 72, 3; 141, 23; 439, 13.

16. Cf. *De Gen. An.* II, 1, 734b, 5-7.

17. **Ambiguous**: The Hebrew translator apparently read *ta'arîḍ* (or *i'tirâḍ*), rather than the *ta'âriḍ* of the Arabic MS, and I believe his reading was correct. *Ta'âriḍ* normally means "contradiction" (as in *FM* 13,12), a meaning which seems inappropriate here (see below, n. 18). See also *LM* XII, Comm. 53 (Arabic: p. 1717, 1-3; Latin: 338 K), where *'i'tirâḍ* and *shukûk* represent the Greek *atopa* and *aporiai* of *Metaphysics* XII, 10, 1075a, 26-27; and the translation of *aporêseien*, in *Physics* II, 8, 198b, 33, as *alladhî yudkhilu al-shakk* (p. 145, 13; Latin: [*LP* II, Text. 78, 77 D] *dubitatio*). Cf. *TAT* IX (Arabic: p. 405, 10; Latin: 101 G) where *dubitationes* translates *i'tirâḍ*. It is also possible that the Hebrew phrase in our text is to be taken as the equivalent of *agônistikon kai eristikon* (see Wolfson, *Crescas*, p. 397).
Assumed: The translation is based on the reading of the Arabic. Here the Hebrew translator was again confused by Arabic orthography: he read a verb from the root *w.dj.d* instead of a verb from the root *'.kh.dh*. In In Arabic, it is a difference of only a few dots.

18. *EDGA* II, cap. 1, 71 L:

> Solvere igitur debemus hoc dubium, quod sic ipsum solvere convenit, declarando scilicet quod id, quod dicitur conditionaliter et cum hypothesi, assumitur hic absolute. Nam, cum dicimus ipsum extrinsecum agere, et non agere, si hoc absolute sumatur, sequentur ex eo priores falsitates. ut si dicamus ipsum foetum generari propter ipsum extrinsecum tantum, sequentur illi priores errores.

This mention of the conditional and the absolute is based on *De Sophisticis Elenchis* V, 166b, 38 ff., and *ibid.*, XXV, 180a, 23 ff. (cf. *ibid.*, VI, 168b, 11-16). The terminology used in this *Question* is in perfect accord with that of *KMH*, *Ḥat'a'â*, pp. 52b f. (Latin: *De Elenchis*, 69 K-70 B). For similar terminology, see, e.g., *Int. De Soph. El.* I, cap. 4, 144 H, and cap. 5, 147 GH; *LM* XII, Comm. 37 (Arabic: p. 1603, 15; Latin: 320 C [*absolute/in comparatione*]), and *ibid.*, VII, Comm. 12 (Arabic: p. 792, 4-5, where *taqyîd* = *ishtirâṭ*; Latin omits). See also Narboni's commentary to *Moreh Nebukim* II, 6 (p. 28):

> By **absolute** he means without a "so to speak"--which would indicate conditionality, or doubt of one's knowledge, or equivocation.

It should be noted that the Greek text of *De Soph. El.* does not offer an etymological equivalent for the Arabic and Hebrew terms translated here as **qualified**.
Aristotle's qualification of the assumption of an external mover is found at *De Gen. An.* II, 1, 734b, 5 ff.

19. Cf. *Question* II, paragraph 6.

20. See *De Gen. An.* II, 1, 734a, 2-4; *EDGA* II, cap. 1, 71 EF; *EP* VIII, p. 150, 3-5 (Arabic). Cf. *De Gen. An.* I, 21, 729b, 21-32. In the Aristotelian system, all motion depends on physical con-

tact between the mover and what is moved.

21. **Disappears**: lit., "is corrupted," "perishes." The Hebrew *yippâsêd* and the Arabic *yublî* = *phtheiretai* (*phtheiresthai* in *De Gen. An.* II, 1, 734a, 9 ff.; cf. *LM* III, Text. 12 [Arabic: p. 235, 15], and Comm. 12 [p. 239, 8]: "...for all the sensible things are perishable [*bâliyya*] <and> corruptible [*fâsida*])." Cf. also *TAT* IV, p. 159 (Arabic: pp. 268, 5-269, 11; Latin: 70 B-E), and van den Bergh's notes *ad loc.*

22. *De Gen. An.* I, 23, 731a, 14-19; cf. *Physics* VIII, 6, 259a, 1 f.; and *IP* VIII, iv, 4, 6 (fol. 126a, 1-2):

> ... for we perceive in the species of animals which reproduce sexually that that which is generated from them is completed and continues to grow when its generator has already perished...

23. The Hebrew translator appears to have had a better reading (*wâlî*) before him than the reading of our Arabic MS (*walî*).

24. *De Gen. An.* II, 1, 733b, 32-734a, 1

25. The Hebrew MSS read: "even if it had a begetter." This is an error on the part of the translator who read *wa-'in* for *wa-'an*.
The argument Averroes presents here does not occur in *De Gen. An.*; rather, it is an interpretation of *De Gen. An.* II, 1, 734a, 5-7.

26. Perhaps this is an interpretation of *De Gen. An.* II, 1, 734a, 13 ff.

27. Cf. *ibid.*, I, 17.

28. See *ibid.*, I, 18; cf. *ibid.*, II, 1, 734a, 33-734b, 3.

29. See *ibid.*, II, 1, 734a, 16-25.

30. **Latency**: The Hebrew term used here commonly means "flame" (*lahaṭ*). But here it is used as the equivalent not of the Arabic *laẓâ*, which also means "flame," but of *laṭâ*, another term for latency. In Arabic orthography, only one dot differentiates the two terms.

31. In his commentary on our text, Narboni distinguishes between two groups of philosophers who maintained that the semen proceeds from the whole body. The first group maintained that the members of the body produce actual members in the semen, but that these members are imperceptible. The second group maintained that there are no actual members of the body in the semen, but that each member of the body contributes to the semen. According to Narboni, Aristotle denied both these opinions and refuted them by ten arguments, some serving to refute both opinions, some peculiar to one of the two opinions. When Averroes says, **We are not able to say of the animal that from each member of the generator there proceeds a similar member [in the semen]**, he is referring to the

opinion of the first group: i.e., actual members in the semen. Narboni continues:

> But we see animals in which the male does not emit semen into the female, and where this emission is not necessary for procreation--much less that it should be necessary that the semen should proceed from all the body, and still less that the semen should contain actual members. Inasmuch as this [last-mentioned] doctrine is the doctrine of one who believes that generation is latent (i.e., that everything exists in everything and is discriminated from the mixture according to what prevails, and that generation consists only in being discriminated from the mixture), Averroes referred to this absurdity when he said, **Furthermore, if the members were generated simultaneously, the embryo would be produced according to latency (i.e. *kumûn*), not according to generation, and the impossibility of all this has already been demonstrated.** He meant: furthermore, if it proceeds from the compound, there is no generation; rather, there is a coming forth from latency, so that what is produced exists before it exists.

Narboni goes on to discuss certain other aspects of latency theory: the role of the agent as reduced to being merely a source of the discriminating motion; the relationship of this theory to the theory of creation; Avicenna's opinion, etc., his discussion consisting for the most part of quotations taken from *LM* XII, Comm. 18 (Arabic: p. 1497, 15 ff.; Latin: 304 E ff.; for a Hebrew text, see Goldstein, "Sources"). Note that in the text from *LM*, *kumûn* is opposed to creation *ex nihilo*.

On **latency**, see also *Physics* I, 4; cf. *LP* I, Comm. 33, 22 BC; *LM* XII, Comm. 8, 296 AB (Arabic: pp. 1440, 13-1441, 8). For a discussion of *kumûn* as treated by Muslim authors, see Wolfson, *Kalam*, pp. 495-517.

32. **Live:** the reading of the Arabic MS, *yu'ayyi<u>sh</u>* (= Gr. *zêi*) is the term used in the Arabic translation of *De Gen. An.* I, 18, 722b, 4 f., which is the source of this passage. The reading of the Hebrew MSS (*hâyâ* instead of *ḥâyâ*) must be corrected accordingly. Cf. *EDGA* I, cap. 20, 63 F: *nam non datur aliquod membrum animalium nisi animatum, i. habens vitam: neque ita se habet, nisi sit aliqua pars corporis.*

33. See *De Gen. An.* I, 18, 722b, 3-5 and 21-24.

34. Narboni *ad loc.* quotes *EDGA* I, cap. 18, 56 DE. Cf. *ibid.*, cap. 20, 63 F-H.

35. Presumably, the heart is the example because of the belief that it developed first in the embryo and that it was also the first principle of the natural body. See *De Gen. An.* II, 1, 734a, 16-25 and 735a, 22-26; *ibid.*, II, 4, 738b, 15-20; *ibid.*, II, 6, 743b, 25 ff.; *EDGA* II, cap. 4, 85 F-86 C; *ibid.*, II, cap. 1, 72 IK.

36. See *De Gen. An.* II, 1, 734a, 25-33; cf. *EDGA* II, cap. 1, 71 I. Narboni comments:

> **The form of all the members would be in the heart, actually.** Averroes means by this that it would be heart and marrow and testicles simultaneously. This is so because neither in art nor nature is the potential produced and generated from what is simply actual, but from the [appropriate] actual. That is, it does not suffice that what is generated should be generated from *any* actual being, such as one member of the body being generated by another which is actual. Rather, both in art and nature, it is necessary that it should be generated by what is actual in so far as it is an actual being of the same species or genus as that generated by it; for the agent must agree either in species or in genus...

Narboni's analysis of the argument agrees with what we find in *EDGA* II, cap.1, 71 HI.

37. See *De Gen. An.* II, 1, 734b, 19-22; cf. *ibid.*, II, 5, 741b, 7 ff.; and *Metaphysics* IX, 8, 1049b, 24-29. A brief recapitulation of many of the arguments used to this point in this *Question* may be found in *EDGA* IV, cap. 3, 118 G-K.

38. This passage is very difficult grammatically in both the Hebrew and Arabic texts, but Averroes is apparently filling in what he perceives to be a gap in Aristotle's argument. Aristotle's analogy between father and artisan, semen and tools, is not completely valid. First, the artisan does not make another like himself. Second, the artisan is in contact with what he is making, either directly or through manipulating his tools, until the product is complete. Thus, it is clear that neither the artisan nor his tools need to be of the same species as the product. For this reason, Averroes emphasizes the semen's specific identity with the father. The semen both generates and is generated, and since it is a medium between father and offspring, it must resemble both.

I have not found any other passages in either Aristotle (although the argument must in some way refer back to *De Gen. An.* II, 1, 734b, 34-735a, 4), or Averroes, in which this is stated explicitly. For Averroes, the closest approximation is *LM* XII, Comm. 24 (Arabic: p. 1529, 4-5; Latin: 309 EF: *Medicina enim aliquo modo est sanitas... et semen hominis homo*). The normal emphasis in both Aristotle and Averroes is on the specific, or generic, identity of father and offspring. See, e.g., *Metaphysics* VII, 8, 1033b, 29 ff.; *LM* VII, Comm. 28, 178 E (Arabic: p. 868, 17-18); *ibid.*, Comm. 31, 180 K and 181 B (Arabic: pp. 881, 1-7 and 882, 13-14); *ibid.*, XII, Comm. 18, 304 K-305 A (Arabic: pp. 1499, 17-1500, 15); *EM* II, 367 B (Arabic: p. 49, 12-17; German: p. 40); *LDC* II, Comm. 42, 126 K.

The words **or genus** belong in the text. It is this formula which extends the theory of generation presented here to cover the case of such things as mules.

39. On semen as only potentially the form of that generated, see *LM* VII, Comm. 31, 180 G (Arabic: p. 879, 12-16):

> Deinde dixit forma enim est in potentia, i. forma enim generati est in semine in potentia, quemadmodum forma artificiati est in artifice in potentia, et illud, ex quo est semen, convenit nomine et ratione cum eo, quod fit ex semine quoquo modo.

40. See n. 13, above.

41. See *De Gen. An.* II, 1, 734b, 19-735a, 4, and I, 22, 730b, 9-23; cf. *ibid.*, I, 21, 729b, 18-21; *EDGA* I, cap. 20, 66 H; *ibid.*, II, cap. 1, 72 BC; *LP* VIII, Comm. 47, 388 HI.

42. **When the proximate agent is already absent, Aristotle thought**: All the Hebrew MSS read: **when the proximate agent is already absent, Aristotle did not** (*lô'*) **think....** The Arabic omits the negative particle, but perhaps the Hebrew ought to be emended: **when its** (*lô*) **proximate agent is ....**

43. For the comparison with automata, see *De Gen. An.* II, 1, 734b, 9-19; cf. *ibid.*, II, 5, 741b, 7-9; *EDGA* II, cap. 1, 72 AB; *TAT* III (Arabic: p. 194, 11-13).
The agent that made them: See *TAT* IV, p. 159 (Arabic: p. 269, 1-3; Latin: 70 C); cf. R. Walzer, *Alfarabi on the Perfect State* (Oxford, 1985), IV, 12, 4, pp. 190-191.

44. See *LM* XII, Comm. 18 (Arabic: p. 1501, 6-15; Latin: 305 B):

> Inasmuch as this semen only causes that through the heat which is in it, and heat *qua* heat does not cause anything but softness and dryness and hardness--not a shape and not a form with soul... (my translation)

Narboni, in his commentary to this *Question* agrees: heat and cold may be responsible for the qualities of softness and hardness, etc., but cannot be responsible for the process that results in flesh and bone. His remarks paraphrase *De Gen. An.* II, 1, 734b, 28-735a, 4. Averroes often speaks of the "heat of the heart" in this connection: e.g., *Colliget* IV, 2, 60 DE and *EDAn* I, p. 5, 12-15. See also n. 58, below.

45. **Form of the art**: this is not a very common phrase, but see *Epitome Parva Naturalia* II (p. 50, 8-9 Blumberg), and *LM* VII, Comm. 23, 173 I (Arabic: p. 845, 16), where Averroes comments on the version of Nicolaus (see Drossaart Lulofs, *Nicolaus*, pp. 139-43). Our text apparently refers to *De Gen. An.* II, 4, 740b, 25-34, where this phrase does not occur (cf. *ibid.*, I, 22, 730b, 8-23; I, 21, 729b, 15-20).

46. See *LM* VII, Comm. 31, 180 FG (Arabic: p. 879, 9-16); *ibid.*, XII, Comm. 13, 299 HI (Arabic: p. 1464, 10-17); *ibid.*, Comm. 18, 305 AB (Arabic: pp. 1500, 15-1501, 3 and p. 1501, 13-16). Cf. *EDAn* I, pp. 4, 12-5, 1.

47. See below, n. 51.

48. See *De Anima* II, 1, 412a, 27-412b, 6 (cf. the paraphrase in ha-Levi, *Khazari* V, 12); *LDAn*. II, Comm. 50 (p. 205, 10-14; cf. *ibid*., Comms. 6 and 7); *EDGA* II, cap. 3, 75 G. Cf. *EM* IV, 368 G (Arabic: p. 56, 2-3; the passage is not found in the German edition).

Narboni explains the comparison with the power of the soul by paraphrasing *LM* XII, Comm. 18, 305 AB (Arabic: p. 1501, 3-14). The fact that this power is not itself soul is explained by reference to Aristotle's definition of soul in *De An*., and by the fact that the organs of the embryo are not themselves actual.

On potential soul in the embryo, see *De Gen. An*. II, 3, 737a, 16-18.

49. **The natures of the powers**: I cannot account for the abrupt switch to the plural. Is Averroes thinking of the powers or faculties of the soul generally, rather than the generative power in particular?

50. Cf. *De Gen. An*. II, 4, 415a, 14-20. On the discussion which follows in our text, see *ibid*., II, 3 (cf. II, 1, 735a, 4-9).

51. See H. Wolfson, "Hallevi and Maimonides on Design, Chance and Necessity," *Proceedings of the American Academy for Jewish Research* XI (1941), p. 142, n. 20 (cf. *TAT* III, pp. 126-127 [Arabic: pp. 211, 8-212, 12; Latin: 58 E-H], and van den Bergh's n. 127.3 to his translation [v. 2, p. 83]).

Narboni identifies **the physicians** with Galen; cf. Galen, *On the Natural Faculties* (London and New York, 1916; tr. A. Brock), I, vi, pp. 25, 22-26, 2. Narboni's commentary refers to, and paraphrases, *EDGA* II, cap. 3 (especially 75 I-76 C). He quotes, briefly, Abu Bakr b. al-Sa'ig (in the *Epistle of Departure*: "In the semen of the animal is a divine substance which is intellect..."). He also refers to what Averroes said at the end of *TAT* (the third discussion in "About the Natural Sciences," pp. 357-359 [Arabic: pp. 577, 9-579, 7; Latin: 145 F-L]), maintaining that it contradicts what Averroes said elsewhere, even in *TAT* itself. Other passages quoted at some length are *TAT* IX, p. 245 (Arabic: pp. 407, 2-408, 5; Latin: 101 L-102 B), and *LM* XII, Comm. 18, 305 C-G (Arabic: pp. 1501, 16-1504, 9). Note also the strange reference to his earlier coreligionists and compatriots, cited by Steinschneider, *HU*, p. 180, n. 542; *Die HSS - Verzeichnisse... Berlin*, II, 1, p. 91.

On the commentary of Narboni, and various aspects of this *Question* not treated here, see H. Tunik Goldstein, "*Dator formarum*: Ibn Rushd, Levi ben Gerson, and Moses ben Joshua of Narbonne," in *Islamic Thought and Culture* (International Institute of Islamic Thought, 1982), pp. 107-121.

52. **Nutritive faculty**: the Arabic reads "nutritive form" (*ṣûra*), but it is clear that the comparison here is between the seminal procreative power and the nutritive power of the soul. (It would not be difficult for the term for "power" [*quwwa*] to be corrupted to *ṣûra*.) See *EDAn* I, p. 5, 5: "the formative power [*quwwa*] is of the genus of the nutritive soul." Note, too, that the nutritive form is also the reproductive power (*De An*. II, 4, 415a, 25 ff.; 416a, 18; 416b, 25; *De Gen. An*. II, 1, 735a, 15-20).

The Hebrew and Arabic texts use the same term (*kôaḥ/quwwa*) whether speaking of powers in semen or powers (faculties) of the soul. The use of term "faculty" for the latter in our translation is intended to clarify the meaning of the text.

53. *De An.* II, 2, 413b, 1-13; 414a, 25-28; cf. *EDAn* I, p. 13,1-10, and *LM* XII, Comm. 18, 304 M (Arabic: p. 1500, 9-10).

54. Cf. *De Gen. An.* II, 4, 740b, 26-741a, 2.

55. See *De An.* II, 4, 416b, 9-25; cf. ha-Levi, *Khazari* V, 12.

56. On the identity of formal and final cause, see *Metaphysics* VIII, 4, 1044b, 1; *De Gen. An.* I, 1, 715a, 4-9; *De Gen. et Corr.* II, 9, 335b, 5-7.

57. See *EDGA* II, cap. 3, 75 K; this passage is paraphrased in Narboni's commentary on the *Question*. But cf. *De Gen. An.* II, 4, 740b, 25-741a, 1, where the nutritive soul and the generator are identified.

58. Cf. *LM* VII, Comm. 31, 181 D-F (Arabic: pp. 883, 15-884, 11), where the same point about heat and organs is made, but the generative power is compared with intellectual powers, not the nutritive power. Also cf. *EDAn* I, pp. 15, 11-16, 5, where (among other things) we are told that the generative power *is* a soul, that its instrument is innate heat, and that there is no difference between this power and the nutritive power except that the generative power makes what is potentially an individual of its species an actual individual, while the nutritive faculty makes only part of an individual.

59. See *De Gen. An.* II, 3, 736b, 14-20. Cf. *ibid.*, II, 1, 735a, 12-13; *ibid.*, II, 3, 736a, 24-32; *De An.* II, 2, 413b, 24-27.

60. **When Aristotle had investigated**: the reading of the Hebrew (*le-mâ še-ḥâqar*) would normally mean "inasmuch as... had investigated." The translator has given us a rather careless version of the Arabic *li-mâ*, as used with the perfect verb.

61. **It appeared**: the Hebrew verb used is *nimṣâ'*, a somewhat surprising choice on the part of the translator . One would expect either *nir'eh* or *nigleh*, the verbs used elsewhere in the translation of these *Questions*.

62. See *De Gen. An.* II, 3, 736b, 21-27; cf. *EDAn* I, p. 5, 7-11.

63. See *De An.* III, 4, 429a, 18-29; *LM* XII, Comm. 17, 302 K (Arabic: p. 1487, 10-13).

64. See *De Gen. An.* II, 3, 736b, 27-29; *EDGA* II, cap. 3, 75 EF.

65. Narboni, *ad loc.*:

> By **it is generated in certain respects**, he means because of the disposition which is dependent on the imaginative forms and the

imaginative soul, and by **it enters from without in certain respects**, he means because of the Active Intellect which is conjoined to this disposition.

See *LM* VII, Comm. 31, 181 KL (Arabic: p. 886, 11-15); *De An.* I, 4, 408b, 18 ff. On the relation between imagination and thinking, see *De An.* I, 1, 403a, 8 f.; II, 3, 414b, 28-415a, 12; III, 3, 427b, 14 ff.; III, 7, 431a, 14 f.

66. Narboni takes this to be a reference to *De An.* Perhaps he had III, 4-8 in mind; possibly, II, 2, 413b, 24 ff. Cf. *LM* XII, Comm. 17, 303 AB (Arabic: p. 1489, 7-12).

67. See *De Gen. An.* II, 3, 736b, 27-29.

68. **Decide**: The Arabic reads *qaḍâ ʿalâ*. Commonly, this means "decide against," a meaning which does not seem possible here.

69. Aristotle is speaking of the semen, or some substance in it, as carrier of the powers that produce the soul.

70. The **body** spoken of here is still the semen.See *De Gen. An.* II, 3, 736b, 29-737a, 7; and *EDGA* II, cap. 3, 75 FG:

> Et cum id ostensum fuerit, supponamus hic veluti per se notum, quod omnis anima materialis communicans alicui corpori composito, omnis nobilior est quatuor elementis, quamvis ex ipsis constet illud corpus: et quemadmodum differunt animae inter se nobilitate et ignobilitate, sic etiam haec nature, quae animam recipit.

On the passage from *De Gen. An.*, and its relation to other aspects of Aristotle's theory of generation, see F. Solmsen, "The Vital Heat, the Inborn Pneuma and the Aether," *Journal of Hellenic Studies* 77 (1959), pp. 119-23.

71. This brief formula of closing, similar to what is found in the Hebrew text of *Questions* I and IV, exists in only one of our Hebrew MSS of this *Question*. The Arabic MS has a longer closing formula, including the customary *salâ* for the Prophet.

QUESTION VII

1. This *Question* is probably identical with the "Appendix to the Seventh and Eighth Books of the *Physics*" (*taʿlîq al-maqâla al-sâbiʿa wa-al-thâmina min al-samâʿ al-ṭibîʿî*) mentioned in the Escurial list (Renan, *Averroès*, p. 464; Steinschneider, *HU*, p. 180). In the Arabic MS of the *Question*, the title is "A Treatise of Abu'l Walîd [Ibn Rushd] Concerning the Seventh and Eighth Books of Aristotle's *Physics*."

In his commentary, Narboni calls this *Question* "the extraordinary treatise" (*ha-ma'amâr ha-niplâ'*) and, at the end of his commentary, "the inclusive treatise" (*ha-ma'amâr ha-kôlêl*; cf. above, *Question* V, n. 1). At the beginning of the commentary, however, he refers to it as the "most special of the treatises" (*ha-meyûḥedet she-be-'iggerôt*). Narboni claims that, for two reasons, his comments will be brief. First, the treatise deals with two books of *Physics* on which he himself has already completed a commentary, so that there is no need to repeat himself (cf. Steinschneider, "Alfarabi," p. 121; *HU*, p. 181). The second reason is to have the commentary agree with the practice of Averroes himself: to be brief and to include only what is indispensable. Thus, Narboni will treat only what Averroes says in this treatise, without bringing forward material from *Physics* which would make it more complete; for the principles discussed in those books of *Physics* ought to be safeguarded on the ground that they lead to the truth about the Prime Mover who is called God.

It will be remarked in the appropriate notes below that this *Question* contains certain passages which appear to be doublets. If Averroes did write alternate versions for some parts of this treatise, and some later editor or copyist included all the versions, this must have been done before our Arabic MS was copied, and before the Hebrew translation was made. Narboni comments on the whole text without, apparently, realizing that doublets exist.

2. *Physics* VII, 1, 241b, 24-242a, 15 (*textus alter*; all references to *Physics* VII, in this *Question*, will be cited from this version).

3. *Ibid.*, VIII, 4

4. *Ibid.*, VII, 1, 242a, 15-243a, 2

5. *Ibid.*, VIII, 5

6. On the problem of the authenticity of *Physics* VII, 1 and its function as viewed by Averroes, see my "Averroes on the Structure and Function of *Physics* VII, 1," *Harry Austryn Wolfson Jubilee Volume* I (Jerusalem, 1965), pp. 335-55. (A reference to this problem in Averroes' commentaries, inadvertently omitted from that article, is *LP* VIII, Comm. 1, 338 F.) Generally, Averroes puts a higher valuation on this chapter than either Alexander (see above, *Question* I, nn. 16 and 26) or Simplicius (*in Physicorum*, pp. 1036-37). Al-Farabi, however, appears to have accepted *Physics* VII as genuine and integral (see A. Birkenmajer, "Eine wiedergefundene Übersetzung Gerhards von Cremona," *Beiträge z. Gesch. Philos. Theol. Mittelalters*, Supplement band 3, especially p. 479).

A large part of this *Question*, however. deals with the difference between the demonstrations in the earlier part of *Physics* VIII and those which occur later in the same book. This was noted by Narboni:

> ...that the particular proof Aristotle employed in the eighth book with reference to the fact that the elements do not move themselves is different from the proof he employed in the seventh book. Similarly, with reference to the eternity of the first thing in motion and its motion, the proofs at the begin-

> ning of the eighth book and those at the end of the book [are different from one another]. For the proof differs, and necessarily so, as is demonstrated in this extraordinary treatise.

7. The Hebrew text of this *Question* is frequently interrupted by the commentary of Narboni. The words **Averroes said** indicate the resumption of the text of the *Question* itself. Henceforth, these words will be omitted in the translation.

8. Aristotle's primary definition of Nature is given in *Physics* II, 1, 192b, 21-23:

> Nature is a source or cause (*archês tinos kai aitias*) of being moved and of being at rest in that to which it belongs primarily, in virtue of itself and not in virtue of a concommitant attribute.

(Cf. the Arabic version [and the marginal comment], p. 79, 7 ff.) The wording of *Physics* II is repeated in *LP* II, Comm. 3, 49 B; *EP* II, p. 5b, 8-10 (Arabic: p. 20, 10-11); *IP* II, 1, fol. 16b, 1-3 (Latin: 442 C).

In our text, however, Nature is described only as **principle**, not "principle and cause." Narboni, *ad loc.*, tries to explain why:

> Averroes undertook the proof of this through what had been demonstrated about the essence of Nature: that it is the principle of motion and that which causes the motion; and that it is that which exists *in* what is moved, [but] is not identical with it.... Inasmuch as matter is cause by way of receptivity, the term "cause" which refers to the matter, apart from the [motive] principle, was mentioned there [*Physics* II] in his expression, "a certain principle and cause." But because the investigation here is peculiar to *motive* cause, Averroes mentions only the "principle"...

Pace Narboni, Averroes appears to be drawing on *Metaphysics* V, 4, 1015a, 13-19 (cf. *LM* II, Comm. 16, 36 A [Arabic: p. 52, 7-14]; *ibid.*, V, Comm. 5, 108 G [Arabic: pp. 514, 16-515, 1]; *ibid.*, IX, Comm. 13, 239 M-240 A [Arabic: p. 1179, 5-15]), where Nature is the essence of things that have in them a "principle"--not a "cause"--of motion.

For other mediaeval versions of the definition of Nature, see Wolfson, *Crescas*, p. 672.

9. *EP* II, p. 5a, 25-27 (Arabic: p. 19, 9-11):

> Consequently, it is evident that the natural bodies are to be distinguished from the artificial bodies. The natural bodies are those which have in themselves a principle of motion and rest, and here I mean by "motion," change; and by "rest," the privation of change.

IP II, i, 1, fol. 16b, 3-4 (Latin: 442 CD): "Here, by 'motion' and 'rest,' he means simply 'change and its privation'."

10. *EM* I, 363 G (Arabic: pp. 34, 19-35, 1; German: p. 27):

> Natura dicitur de quolibet genere quatuor transmutationum, scilicet generationis et corruptionis, et motus translationis, et augmenti, et alterationis.

11. Narboni, *ad loc.*:

> Inasmuch as this *Question* is concerned with all things in motion (i.e., that they have a mover other than themselves, either in them or with them, even that which is really in motion primarily of itself, namely, the outermost sphere), he explained that that motion which is mentioned here in the definition of Nature includes all the motions, and the principle mentioned includes all the principles of motion, whether inanimate, animate, or rational.

Note, however, that when Averroes speaks of the Prime Mover (below, paragraph 42), he uses not the term "nature," but the term "force" (= a source of motion in something else, or in itself *qua* other). Cf. *De Caelo* III, 2, 301b, 17 ff., and Simplicius, *ad loc.*

12. *Physics* II, 1, 192b, 27-29; cf. *Metaphysics* XII, 3, 1070a, 7-8. In Averroes, see *LP* II, Comm. 2, 48 IK; *LM* XII, Comm. 13, 298 M-299 A (Arabic: pp. 1459, 15-1460, 1). Cf. Alexander, *De Prin.*, pp. 253-54.

13. Cf. *LP* VIII, Comm. 60, 400 A. Narboni's comment on our text is:

> Then, by an inductive consideration of things in motion, he demonstrated the second point in the definition of Nature: that it is a principle *in* something moved, not the thing in motion itself in such a way that the mover is identical with what is moved.

14. The second kind (elements) is not mentioned until paragraph 6.

15. **Its need for the mover is clear:** This was apparently the reading of the Arabic exemplar of two of our Hebrew MSS (a third MS omits): *wa-hadhâ bayyana iḥtâdjuhu 'ilâ muḥarrik.* Narboni accepted the reading of the Hebrew MSS, but it seems clear that the reading of the extant Arabic MS is to be preferred: "and in this case the difference between the mover..." (*wa-hadhâ bayyana mughâyar al-muḥarrik.*) The latter agrees with *Physics* VIII, 4, 254b, 31-33 (and the reading of the Arabic version of that passage [p. 836, 5-6]), and Averroes' own usage in such passages as *LP* VIII, Comm. 28, 366 EF, and *IP* VII, i (see below, n. 16).

16. **For it comes to rest when the soul no longer exists:** I have retained this phrase, omitted in the Arabic text, on the basis of the following passage from *IP* VII, i, fol. 100b, 20-101a, 9:

> I say that everything moved in place without some external mover must be in motion due to some mover distinct from that which is

> moved, for it is obvious that those things which are moved by external movers are moved by a mover distinct from that which is moved. The only things about which doubt occurs are the things moved in place without an external mover, and in particular, the simple bodies [elements]... for one might think that these bodies are moved by themselves, and that the mover in them is identical with what is moved. As for the animal, the difference between what is moved in it and the mover is readily apparent, because it is self-evident that the soul causes motion in it, while the body is what is moved, for its motion ceases to exist when the soul no longer exists.

Cf. Elias Cretensis, *op. cit.*, fol. 154, col. 1 AB. Narboni comments on our text:

> He means that it comes to rest when the soul no longer exists even if the soul is a part of the totality compounded from what is in motion and the mover. The soul is different from what is moved, namely, the corporeality; and it is different from the compound as the part is different from the whole.

17. See *LP* II, Comm. 1, 48 F.

18. Note that here, as in *Physics* VIII, 9, 265b, 32-34, although the reference is clearly to Plato, his name is not mentioned. Cf. *Question* I, n. 15, and below, paragraph 44, and n. 172.

19. I have combined the reading of the Hebrew MSS with that of the Arabic, as required by the context; cf. *LP* VII, Comm. 3, 308 H [*sic*].

20. See *Physics* VIII, 5, 256a, 4-256b, 3; cf. *ibid.*, VII, 1, 242a, 16-243a, 2.

21. The text is difficult. I take this to be a paraphrase of Aristotle's discussions at *Physics* VIII, 5, 256b, 3ff., and *ibid.*, 6, 259b, 16 ff (cf. Abu'l-Faraj, *ad loc.* [pp. 871, 9-872, 9]). The point is that no moved mover, even if it is only moved accidentally, can cause continuous motion. See *LP* VIII, Comm. 36, 375 H-K.

22. *IP* VIII, iv, 4, 2, fol. 120b, 3-11:

> Because it is evident that these two kinds of things in motion are moved by movers other than themselves, difficulty occurs only with respect to the simple bodies, that is, the heavy and light bodies..., for it is clear that when these bodies are moved violently, they are moved by a mover other than themselves.... The difficulty concerning them occurs when they are moved with their natural motion, for then it is thought... that they are moved by themselves, and that in them the mover is identical with what is moved.

Cf. *LP* VIII, Comm. 28, 366 H-K; *ibid.*, II, Comm. 3, 49 B-D. On the problem of the motion of the elements, see Pines, "Refutation," pp. 36 ff.; Wolfson, *Crescas*, pp. 672-75.

23. Later in this treatise, Averroes demonstrates that **the mover which is in addition to their substance** is an external mover (paragraph 23). Here, however, Narboni explains that this mover is the form of the elements: "... they are moved by a mover other than themselves, or something in addition to their corporeal substance, namely, their form." Narboni's explanation agrees with *EP* VII (p. 115, 8-12 Arabic), *EDC* IV, p. 68, 4-6, and *LP* IV, Comm. 71, 161 K; cf. *TAT* XIV, p. 287 (Arabic: pp. 471, 15-472, 7; Latin: 117 DE).

24. *Physics* VII, 1, 241b, 24-242a, 15. (On the proof given there as against the proof in Book VIII, see *LP* VIII, Comm. 28, 366 B-D.) In reproducing this argument, Averroes reverses the order of Aristotle's proof. For another analysis of *Physics* VII, 1, see above, *Question* I. *EP* VII, 1, pp. 34b, 27-35a, 1 (Arabic: p. 113, 2-5)

> Aristotle employed three propositions at the beginning of the book [Arabic: at the beginning of the seventh book] to demonstrate that everything in motion has a mover, and that nothing exists which is in motion of itself, that is, so that what is moved is identical with the mover, as might be thought with respect to earth and water and the bodies which are in motion without an external mover.

(Cf. Elias Cretensis, *op. cit.*, fol. 153, col. 1 H.)

25. **Which is something additional in the subject**: all the Hebrew MSS contain these words. However, the words are missing in both the Arabic MS and Narboni's lemma (although they do occur in his commentary on this text.) We may have here an example of a gloss which has been incorporated in the text.

26. Note that the propositions which follow are not identical with those found in *EP* VII (p. 113, 5-8 Arabic).

27. *Physics* VI, 4, 234b, 10-20; *ibid.*, 10, 240b, 8-241a, 16. Actually, the proposition is true of what is moved *per se*. Cf. *EP* VII, p. 113, 5-6 (Arabic): "... that everything moved *per se* and primarily is divisible, possessing parts."

28. **Is continuously divisible**: the Arabic reads "is not continuously divisible" (cf. above, paragraph 6, where the Arabic reads: "indivisible", while the Hebrew reads: **inanimate**). Because the first two chapters of *Physics* VI do demonstrate that every body is continuously divisible, I have followed the reading of the Hebrew. The Arabic reading, however, may not be a scribal error, but may refer to the contention that while body *qua* body is continuously divisible, a particular kind of body is not. Cf. Narboni, *ad loc.*:

> **By inasmuch as it has already been demonstrated that every body**

> **is continuously divisible**, Averroes means in so far as it is simply a body, not in so far as it is a certain kind of body, such as fire or air, for example.

In his commentary to *Moreh Nebukim* I, 73 (Proposition I), Narboni goes into more detail:

> You already know that continuous quantity is infinitely divisible, according to the doctrine of the philosophers. They do not mean by this that a part of fire, for example, may be infinitely divided and still remain fire. This would be impossible, for every natural body has a limited minimum and maximum size, and if it passes that limit [in either direction], its form is lost: for then it is out of proportion to its subject. What the philosophers mean is that we ought to consider the magnitude abstracted from the [specific] form. Furthermore, they do not mean that we can divide a body infinitely, in actuality, but only in thought...

See also below, paragraph 6; and *Question* I, n. 6. Cf. *Question* VIII, paragraph 10; *EP* VII (p. 113, 9-11 Arabic).

29. See *LP* VII, Comm. 2, 307 I. Cf. *LP* VIII, Comm. 44, 384 G-K; *ibid.*, Comm. 62, 401 I; *IDC* III, 315 G.

30. Cf. *EP* VII, pp. 114, 1-14 (Arabic).

31. See *Question* I, n. 27.

32. See below, paragraph 11, and n. 43. In *Physics* VII, 1, Aristotle does not explicitly conclude that the first mover must be incorporeal. However, this is clearly the implication of *EP* VII (pp. 114, 15-115, 12 Arabic). Cf. Simplicius, *in Physicorum*, pp. 1040, 26-1041, 3.

33. *Physics* VII, 1, 242a, 15-243a, 2. Averroes telescopes the two arguments found there.

34. The pile-up of explanatory phrases which follows may, again, indicate a gloss (or glosses) incorporated in the text. Cf. below, paragraph 11.

35. See *LP* VII, Comm. 4, 309 AB.

36. *EP* VII, p. 36a, 20-25 (Arabic: p. 117, 4-8):

> Then let us assume a certain mover B, and B causes motion in so far as it is moved because of C, and C causes motion in so far as it is moved because of D, and D [causes motion in so far as it is moved] because of E, and so on to infinity. If, however, we assume this, it follows that an infinite motion exists in a finite time. But how does this follow? Because that which causes motion in place, if it is external [to that in which it

> causes motion], causes motion only when it is in contact with that which is moved.

37. I.e., assuming an infinite number of such reciprocally moved bodies. See below, paragraph 10.

38. *EP* VII, pp. 36a, 28-36b, 2 (Arabic: p. 117, 11-14):

> ... then it is clear that they form a magnitude which is one by contiguity, and that their motion is simultaneous, as is the motion of the parts of one continuous magnitude. Because the motion which is one belongs only to one thing in motion, it is clear that *this* motion is one, for it belongs to something in motion which is one by contiguity.

39. Both the Hebrew and Arabic texts appear to say that the motion is greater than the magnitude moved with that motion, which cannot be correct. Rather, in the Hebrew, we should read: *godlâh mê-ha-gôdel ha-mitnô'êa' bâh*. See *LP* VII, Comm. 8, 311 BC.

40. *EP* VII, p. 36b, 2-7 (Arabic: p. 117, 14-118, 2):

> Then if we assume these moved movers are infinite [in number], it follows that the magnitude compounded from them must be infinite. If we assume this magnitude, in so far as it is a magnitude, to move a finite distance in a finite time, there must be an infinite motion in a finite time; for this motion is in an infinite magnitude.

41. *Physics* VI, 7, 238a, 20-238b, 22

42. I.e., it moves itself through an internal principle of motion which is different from what is moved.

43. Aristotle himself concluded from the argument of *Physics* VII, 1, only that there must be something moved primarily (*textus alter*), or that "there must be a first movent and a first moved" (version *alpha*). Simplicius (*op. cit.*, p. 1047, 15-16) apparently read the latter text as: "there must be an unmoved (*mê kinoumenon*) first movent." Abu'l-Faraj (p. 740, 6-7) offers a similar interpretation: *Physics* VII, 1 proves (1) that everything in motion has a mover; and (2) the series of moved movers ends with a mover which is not in something moved. Averroes' conclusion, that the mover must be incorporeal, has been reached with the aid of material from *Physics* VIII.

44. *IP* VIII, iv, 4, 5, fol. 125a, 6-9:

> Hence it is evident that the mover in it is not a body; for if it were a body, it would suffer motion when it causes motion, that is, it, too, would require a mover. If this mover, too, were a body, what was necessary for the first[-mentioned] mover

would also be necessary for this mover, and this would proceed to infinity.

45. The reference is plainly to *Physics* VIII, 1, 251a, 9-10, but the wording of the abbreviated definition given here seems rather to reflect *Physics* III, 1, 201a, 9-15 (cf. *ibid.*, 2, 202a, 7-8; *ibid.*, 1, 201b, 4-5). The Hebrew term used here (*šelêmût*) is the precise equivalent of the Arabic *kamâl*, the term used in the Arabic version of *Physics* III, where it translates the Greek *entelecheia*. The commentary to the Arabic version explains that *kamâl* is equivalent to *fi'l*, the term used in the translation of *Physics* VIII, where it represents the Greek *energeian* (actuality or actualization). It is the latter terminology that is reflected in the abbreviated definition found in *IP* VIII, ii, 1 (fol. 111b, 16-17: "We say that it had been said previously that motion is the actualization of that whose nature it is to be moved."

46. *Physics* III, 3, 202a, 13 ff.; cf. *Metaphysics* XI, 9, 1066a, 26-34. See also *LP* III, Comm. 18, 92 G.

47. *IP* VIII, ii, 2, fol. 112a, 21-23:

> In general, when that which is moved with the primary motion is assumed to be generated, it follows that the motion which was assumed to be primary is not primary. Furthermore, it is impossible for that motion of generation to be primary, for generation is consequent to another motion.

48. I take this to mean that if the mover and that moved are assumed to be eternal, as they are here, some change must have occurred previously in one or both of them (cf. *LP* VIII, Comm. 9, 345 AB); if either of them is not eternal, it must have been generated, and this, too, would contradict the initial assumption. Cf. Alexander, *De Prin.*, p. 263, 17 ff.

49. Cf. *LP* VIII, Comm. 9, 345 I-M.

50. **In time**: I read *bizman*, with the Hebrew, on the basis of *Physics* VIII, 1, 251b, 9-11. Cf. Abu'l-Faraj *ad loc.*, p. 809, 10.

51. I.e., the motion before the posited motion must be in the mover and must be the cause of the mover's causing motion.

52. I follow the reading of the Hebrew text which better agrees with paragraph 13, above.

53. See *Question* III, n. 2.

54. **Accidents** (*mim-miqrê*): It is not clear to me whether this means "on the basis of the accidents," or "it will be demonstrated that one of the accidents...".

55. **That which is set in motion of itself**: here, this refers to ani-

mals; see below, paragraphs 28-29. On the Kalâm position and Averroes' response to it, cf. e.g., *LP* VIII, Comm. 15, 349 IK.

56. John's name is spelled here in a rather unusual way. Only Hebrew MS. Mun. 31 has the common *Yaḥya*. The remaining Hebrew MSS read *Yaḥnî* (*Yôḥanî?*) which is a transliteration of the name as found in our Arabic MS (*Yûḥanî*; in paragraph 47, *Yûḥannâ*); see *Tafsîr Mâ Ba'd aṭ-Ṭabî'at*, v. III, Index E, p. (305), no. 277.

On Philoponus, see now R. Sorabji (ed.), *Philoponus and the Rejection of Aristotelian Science* (London, 1987). On his influence on Muslims and Jews, see Steinschneider, "Alfarabi," pp. 152-76; H. Davidson, "John Philoponus," pp. 357-91; *Idem*, *Proofs*, especially pp. 86-117; J. Kraemer, "A Lost Passage from Philoponus' *Contra Aristotelem* in Arabic Translation," *JAOS* 85 (1965), pp. 318-27; M. Mahdi, "Alfarabi Against Philoponus," *Journal of Near Eastern Studies* 26 (1967), pp. 233-60; *Idem*, "The Arabic Text of Alfarabi's *Against John the Grammarian*," *Medieval and Middle Eastern Studies in Honor of A.S. Atiya* (Leiden, 1972), pp. 269-84; Pines, "Summary."

57. On this lost work, see Steinschneider, *op. cit.*, pp. 119-23; cf. B. Lewin, "*Muḥdath*," pp. 85-6; S. Pines, Translator's Introduction to Maimonides' *Guide* (Chicago, 1963), pp. lxxxv-lxxxvi.

58. I.e., the potentiality for the motion must itself have been created through a motion.

EP VIII, p. 40b, 11-18 (Arabic: pp. 134, 7-135, 2):

> Moreover, one who raises this objection against Aristotle, as al-Farabi did in his book *On the Mutable Existences*, and as others who followed him did, is one who says that Aristotle's intention at this point was only to demonstrate that every motion has motion before it, and that it was for this reason alone that Aristotle introduced the definition of motion.... John Philoponus before them had already noticed this objection against Aristotle, and he undertook to refute Aristotle because he [Philoponus] assumed that there was motion before motion essentially.

Averroes' knowledge of Philoponus is limited. He does not appear to have known of Philoponus' denial of a fifth essence (I have found no explicit statement in his writings), and therefore does not recognize the full significance of Philoponus' discussion of the definition of motion. See Simplicius, *in Physicorum*, pp. 1129, 28-1131, 7 (now conveniently translated and explicated by C. Wildberg in Philoponus, *Against Aristotle, on the Eternity of the World* [London, 1987], pp. 123-24).

Simplicius reports the argument from infinite regress presented here, with a number of variations, beginning at *in Physicorum*, p. 1178, 5 (tr. Wildberg, pp. 144-46).

59. The Hebrew text differs verbally, but not in sense, from the Arabic. The translator apparently misread *lâ 'an* as *li'an*.

60. Cf. *Question* IV, and *LP* VIII, Comm. 45, 385 EF.

61. See *Question* IV, n. 7.

62. See *Question* IV; Philoponus, *in Physicorum*, pp. 428, 23-430, 10; Pines, "Arabic Summary," especially pp. 330-37 (cf. pp. 347-52); Wolfson, *Kalam*, pp. 410-15.

63. Narboni *ad loc.* quotes from *IP* (VIII, iii, 1, fol. 116a, 19 ff.):

> There is a fourth objection, well-known in our time, namely, that if there are contiguous or [merely] successive motions, it follows that the last motion alluded to (at the time of this treatise, for example) cannot exist unless infinite motions [have existed]. But in what has no end, there is no beginning, and if a first does not exist, neither does the last. The solution of this problem is easy according to the principles already laid down: for the circular locomotions [of the heavens] are one, continuous motion, not many actual successive motions, as they thought; and this had already been demonstrated in what was said before with regard to the conditions of the one motion. But it is impossible for plural successive motions to be infinite, for the non-existence of the prior motion is a condition for the existence of the posterior, and this being so, the non-existence of the prior is a cause for the posterior. Further, since a first cause does not exist, a "last" cannot exist. But this follows only if the non-existence of the prior in a subject is an essential cause for the existence of the posterior. Non-existence, however, is only an accidental cause for generation, as was demonstrated at the beginning of this [eighth] book, and accidental causes can be infinite and no falsehood results from assuming them so. In this way one can resolve the doubts about the animal and the plant which procreate sexually: namely, that the semen is only an accidental cause for the embryo, and similarly, putting forth seed is only an accidental cause for fruit. It was already proved in the *Book of Animals* that the mover in the generation of procreative animals is not the semen.... But the semen... is an instrument for the primary mover, and it is not impossible that the one, eternal mover should cause motion, by means of infinite instruments, in moved things which are infinite. From all these things, we arrive at the resolution of this objection.

See also *Question* IV, nn. 5 and 6.

64. See *LP* VIII, Comm. 1, 339 A-D; *ibid.*, Comm. 9, 345 C-F; *EP* VIII, p. 40b, 18-25 (Arabic: p. 135, 3-8):

> A great difficulty has occurred to the philosophers of our time [Arabic: of our creed] concerning this, namely, the one Alfarabi was forced to posit in this book of his called *On the Mutable Existences*: for there he endeavored to investigate in what

> way it is possible that there should be motion before motion, and to distinguish the impossible in this respect from the necessary. This was because he thought that if there is no necessary species, what was assumed in the definition of motion is absurd; for he thought that Aristotle introduced the definition of motion [here] only in order to demonstrate that before every motion there is motion. All this, however, is a false opinion.

Cf. Steinschneider, *op. cit.*, p. 136; Davidson, "John Philoponus," pp. 359-60.

65. *LP* VIII, Comm. 4, 341 IK:

> Dico secundum hanc expositionem intellexit Alpharabius et alii hoc capitulum, scilicet quod induxit definitionem motus ad declarandum potentiam esse ante actum. et hoc non est proprium motui, secundum quod est motus, sed est proprium novo facto, secundum quod est novum factum, scilicet ut potentia, et posse novi praecedat ipsum secundum tempus.

(Cf. Worms, "Lehre," p. 24.)
Is it possible that Averroes uses the phrase "novo facto" here, and in our text, because he is anticipating Philoponus' distinction between generation and motion in connection with the problem of motion of the elements? See Simplicius, *in Physicorum*, pp. 1133, 30-1134, 4.

66. The eternal motion is circular, and therefore cannot be the "opposite" of the rectilinear motion of the elements. See *De Caelo* I, 4. On the distinction, see *TAT* XIV.

67. **In virtue of themselves**: the Hebrew term *be'aṣmûtâh* is ambiguous; it may refer to the eternal motion (so Steinschneider, *op. cit.*, p. 121), or to the motion of the elements. I take it in the latter sense, following Narboni *ad loc.*, who apparently had a different reading (*be'aṣmâm* or *mê'aṣmûtâm*). The reading of the Arabic MS (*min <u>dh</u>âtihâ*) is equally ambiguous.

68. See Davidson, *op. cit.*, p. 360, n. 24; E. Evrard, "Les convictions religeuses de Jean Philopon," *Academie royale de Belgique, Bulletin de la Classe de Lettres*, 5e série, XXXIX (1953), pp. 309, 312, 313; *TAT* XIV, p. 287 (Arabic: p. 472, 10-12; Latin: 117 EF).

69. Both the text and the argument of the first half of this paragraph are difficult. I can make no sense of the Hebrew text beginning with **it appeared to him**, and I have emended the text on the basis of the Arabic, conjecturing that the translator misread *bi'an* as *bi'anna*.
Having reported, and refuted, Philoponus' interpretation of the meaning and purpose of the definition of motion in *Physics* VIII, 1 (above, paragraphs 15-17), Averroes now turns to Philoponus' attack on the validity of that definition. Unfortunately, Averroes reports the arguments in such a way as to make them almost unintelligible.
He tells us that Philoponus believed that in the case of the natural

motion of the elements, there was no temporally prior potentiality (cf. *LP* VIII, Comm. 4, 341 AB; that Philoponus did hold this view of the motion of the elements seems clear from Simplicius [*in Physicorum*, p. 1133. 23 ff.; tr. Wildberg, pp. 125-26]). Averroes then turns immediately to al-Farabi: because al-Farabi accepted Philoponus' interpretation of the definition of motion, and because he agreed that there is no apparent potentiality temporally prior to the motion of the elements, he was forced to conclude that the potentiality for that motion resided in the body from which the elements are generated.

All this appears to make very little sense, for the motion of the elements is not the issue at hand.

What Averroes does not tell us is that he has reported only half of Philoponus' argument. Simplicius tells us (*op. cit.*, pp. 1130, 15-1131, 29) that Philoponus contended that the definition of motion ought to apply to all cases of motion. But Aristotle's definition does not appear to cover the motion of the elements. Therefore, it must have been Aristotle's intention to establish the existence of an entirely different kind of motion in the heavenly sphere. But in that case, the potentiality in question must reside in something. As Philoponus interprets Aristotle, the potentiality for the motion of the heavenly sphere must reside in the sphere, and so the sphere must be temporally prior to its motion. This means that the motion of the heavens cannot be eternal. (Averroes may allude to this argument in paragraph 18 of our text.)

Averroes also does not tell us that Philoponus himself had considered and rejected al-Farabi's solution of the problem of the motion of the elements (Simplicius, *op. cit.*, pp. 1133, 28 ff.), distinguishing between the generation of the elements and their motion. (Again, we may have an allusion to this in paragraph 18.) Finally, the reference to al-Farabi's solution of the problem is puzzling. In the context of the discussion here, it seems clear that Averroes disapproves of this solution. Yet he himself, with certain qualifications, agreed that the potentiality for the natural motions of the elements exists in that from which they are generated, He said this not only in *LP* VIII, Comm. 4, where he gives al-Farabi's understanding of the text, but also *ibid.*, Comm. 32, 370 C and F-H, and *LDC* IV, Comm. 22, 249 BC (where al-Farabi is quoted approvingly), and *ibid.*, Comm. 25, 253 E. It should also be noted that at *LDC* IV, Comm. 24, 252 IK, Averroes attributes this view to John Philoponus himself. It is clear from Simplicius (*loc. cit.*), however, that Philoponus assumed this solution only for the purpose of showing how many absurdities would follow from it, thus impugning the definition of motion.

We must conclude either that Averroes attacks al-Farabi here because the latter arrived at the solution of the problem of the motion of the elements through a mistaken interpretation of the definition of motion, and because such an interpretation would make it impossible to accept the eternal motion of the sphere; or that this *Question* indicates a real change in Averroes' approach to the problem of the motion of the elements.

70. See *LP* III, Comm. 18, 92 G; *ibid.*, V, Comm. 3, 208 E; *ibid.*, VIII, Comm. 4, 341 K; *EP* VIII, pp. 129, 7-130, 1, and 130, 6-7 (Arabic).

71. See *LP* VIII, Comm. 45, 385 EF; *TAT* I, p. 39 (Arabic: p. 68, 14; Latin: 28 A).

72. See *LP* VIII, Comm. 4, 341 KL. See also Steinschneider, *op. cit.*, pp. 121-22, for a different translation of paragraph 17 and the first part of paragraph 18.

73. I.e., in created things. Cf. *LP* VIII, Comm. 6, 342 C, where the potentiality precedes both that which is moved and the mover.

74. *Physics* VIII, 1, 251b, 10-28

75. Narboni, *ad loc.*, explains that the argument from time is self-evident because the instant is the end of the past and the beginning of the future. Thus, there must always have been time before any given instant, and time is a consequence of motion. This explanation is very close to what we find in *EP* VIII, p. 136, 2-8 (Arabic). Cf. also Alexander, *De Prin.*, p. 264.

76. *Physics* VIII, 2; Averroes, reversing Aristotle's order, takes up the motion of animals first. The motion of the elements is discussed beginning with paragraph 21, below.

77. The reference is to *Physics* VIII, 1, 252b, 5-6.

78. *Physics* VIII, 2, 252b, 17-24. Cf. *LP* VIII, Comm. 17, 353 B-D; and *IP* VIII, iii, 1, fol. 116a, 11-14:

> The third supposed objection--and it is the most persuasive of them--is that we perceive animate beings to be set in motion after they have been at rest without having an external mover, and without any change preceding them, either in them or external to them.

After commenting on the difficulty presented by the case of animals, Narboni (*ad loc.*) cites a lemma not found in either the Arabic MS or the Hebrew MSS:

> When Averroes says, **And he relates what the part of the parts of existing things they sought in this investigation is**, he means what is moved at one time and is at rest at another, and what includes [*yiklôl*; a reference to the sphere?] .

79. **Thesis**: *mebuqqâš*. On this term, see Wolfson, *Crescas*, p. 457, n. 81. The "thesis" is still the eternity of motion; cf. *LP* VIII, Comm. 27, 364 GH.

80. *Physics* VIII, 3

81. *LP* VIII, Comm. 53, 394 G-I:

> Et dixit. Et apparet etiam ex hoc, etc. idest ex praedicitis

> igitur apparet illud, in quo dubitabamus primo, et est, quod non omnes res moventur, neque omnes quiescunt. et causa in hoc est, quòd ista etiam non dividuntur in duos modos tantum, quorum unus est motum semper, et alius quiescens semper sed sunt hic aliqua, quae moventur quandoque, et quiescunt quandoque, et aliqua, quae moventur semper, et quiescunt semper. Et quòd causa in toto declarat est nunc ex hoc sermone. et est, quod quaedam moventur à motore aeterno, quod non movetur, et ideo ista moventur semper: et quaedam moventur à motore moto, et ideo contingit etiam ut ista moveantur quandoque, et quandoque quiescant: et quaedam nunquam moventur, et est motor aeternus, de quo diximus esse motorem uno motu tantum.

See also *LP* VIII, Comm. 19, 354 HI. On the relation of this argument to the earlier arguments in Books VII and VIII, see *ibid.*, Comm. 27, 364 F-K.

82. On the basis of *Physics* VIII, 3, 254b, 4-6, I take this to refer to the position that there are actually three classes of beings, not just two; see below, paragraph 39. Note, however, that in *IP* VIII, iv, 4, 6, fol. 126b, 10-17, Averroes says that the intention is to demonstrate that some things are moved eternally and others are eternally at rest:

> ... for inasmuch as it is evident that, of the things that exist, some are moved at one time and at rest at another--and for this reason we said that not all things are in motion nor are all of them at rest; neither are all of those which are in motion in motion eternally, nor all of those at rest at rest eternally --and inasmuch as it was our intention to demonstrate the second part, namely, that some of the things in motion are in motion eternally, and some of the things at rest are at rest eternally: we laid down the principles of the speculation about this when we demonstrated first that everything in motion has a mover, and that the mover must inevitably be one of three things...

If this is what Averroes means to say in our text as well, then we have here an anticipation of *Physics* VIII, 6, 259a, 27-29.

83. The reading of the Arabic MS is "to decide."

84. **The heavy and light [bodies]**: Henceforth, this will be translated simply as "the elements." On what follows, cf. *LP* VIII, Comm. 28.

85. *Physics* VIII, 4; cf. *EP* VIII, p. 138, 2-5 (Arabic):

> Thus, it remained for him to investigate here into which class of things in motion the elements fall: the class of those things moved by something external to them, or the class of those things moved by themselves.

86. I.e., the elements and animals.

87. *EP* VIII, p. 138, 5-6 (Arabic):

> When he had investigated them here, it was clear to him that they are not to be numbered in the genus of those things which are in motion of themselves, except accidentally.

See also *LP* VIII, Comm. 31, 369 A, and cf. *LDC* III, Comm. 28, 198 L.

88. The explanation of Aristotle's purpose which follows in our text is of quite a different order from the explanations given in the various commentaries on *Physics*.

The text of *EP* VIII (Arabic: pp. 140, 13-141, 7) is difficult, as may be seen from the Hebrew translation (p. 43b, 12 ff.). It states:

> Thus, it is already evident from this that Aristotle did not [merely] return for a second time to the demonstration that there is something in motion primarily, nor could he have been satisfied [to stop] with the demonstration employed in the seventh book. Similarly, it is also clear that he used [that demonstration] here potentially, as [he used] the principle which was [also] demonstrated in the seventh book: that everything in motion has a mover. For what is evident here [in Book VII], that that these simple bodies are moved by what is external to them--and it is this which occasions a difficulty concerning them: whether the mover in them is identical with what is moved --is not sufficient to demonstrate that everything in motion has a mover, if their external mover is a mover by accident. And [it is evident that] what Aristotle uses for proof of this in this [eighth] book is only by way of additional proof and testimony, as was his habit, [but] is not a sufficient demonstration in itself. Thus, Themistius erred in more than one respect: first, because he thought that the inductive proof which occurs here was sufficient to demonstrate that everything in motion has a mover; second, because he thought that what is demonstrated here about that which is in motion primarily is identical with what was demonstrated at the beginning of the seventh book, so that he transferred the demonstration used there to this place.

IP VIII, iv, 4, 2, fol. 122a, 1-7:

> Inasmuch as it has already been demonstrated that these bodies are not in motion of themselves but are only in motion due to something external to them, and doubt ocurred [only] as to whether everything in motion is moved by something, it is [now] clear by induction that everything in motion has a mover. But this had already been proved by a veracious demonstration at the beginning of the seventh book. For this reason, what is to be thought about the use of the inductive proof here is that it is only by way of additional evidence, not that it is sufficient in itself, as Themistius thought...

LP VIII, Comm. 33, 372 B-F:

> ...Et videtur mihi quod ipse intendit per hanc declarationem hic dissolvere quaestionem accidentem in propositione, dicente quod omne motum habet motorem ex motis simplicibus. ista enim apparent moveri per se, cum fuerint extra sua loca naturalia, et sunt simplicia non divisibilia, scilicet quod motor in eis est ipsum motum. ista igitur, quia faciunt dubitare in hac quaestione, secundum quod Aristo. dixit in principio Septimi, ideo incoepit illic demonstrare hanc propositionem, scilicet quod omne motum habet motorem, sed non dissolvit illic quaestionem accidentem in eis ex his simplicibus. in hoc autem loco dissolvit illam quaestionem, et declaravit quod ista moventur ab extrinseco, et quod non moventur ex se, nisi per accidens.... Et intendebat quod ista propositio declarata est hic perfecte, cum quaestio, quae accidit in ea ex corporibus gravibus, et levibus, sit dissoluta: et demonstratio facta sit de ipsa in principio Septimi: non, quia inductio, qua utitur hic, sit sufficiens in hoc. inductio enim in qua non percipitur quod praedicatio est essentialis, non dat certitudinem naturalem, licet inducantur in ea omnia particularia.

On the argument of *Physics* VIII not being more persuasive than that of *Physics* VII, see below, paragraph 27; Simplicius, *op. cit.*, p. 1042, 8-9 (cf. pp. 1036-37).

89. I.e., the proof in *Physics* VIII, 1.

90. Here, **the two of them** seems clearly to refer to the elements, i.e., the heavy and the light. Narboni (*ad loc.*), however, says that the sentence, up to the words **by a mover external to them**, means that this is one proof appropriate to both animals and the elements.

91. *LP* VIII, Comm. 33, 372 CD:

> ... ideo incoepit illic [in principio Septimi] demonstrare hanc propositionem, scilicet quod omne motum habet motorem, sed non dissolvit illic quaestionem accidentem in eis ex his simplicibus. in hoc autem loco dissolvit illam quaestionem, et declaravit quod ista moventur ab extrinseco, et quod non moventur ex se, nisi per accidens.

See also *ibid.*, V, Comm. 18, 221 C; *LDC* IV, Comm. 24, 251 M-252 A.

92. *LP* VIII, Comm. 17, 353 G:

> ... de motu enim naturaliter apparet quod non moventur, nisi generentur in locis extraneis, aut patiantur violente quietem, et post recedet violentans. et universaliter motor est generans in eis, aut auferens violentiam, ut declarabitur post. Et ideo dixit sed ista semper moventur ab aliquo extrinseco. vivum autem dicitur movere se: et utrunque in rei veritate, ut post

declarabitur, indiget motore extrinseco: sed est latentius in habentibus anima.

See also *TAT* XIV, p. 287 (Arabic: p. 472, 10-12; Latin: 117 EF); *LDC* III, Comm. 28, 198 G ff.

93. See *De Caelo* IV, 3; cf. *De Gen. et Corr*. II, 8, 335a, 17-22; *LDC* IV, Comm. 24, 252 F; *ibid*., Comm. 25, 253 EF; *LP* VIII, Comm. 4, 340 M-341 A.

94. See Pines, "Refutation," pp. 37-40. The bracketed words in my translation can be justified on the basis of the following passages: *EDC* IV, p. 79, 8-10:

> The place of the part and the place of the whole, as has been said, is one; and its entelechy is the thing by which the whole is perfected: namely, the place which fulfills the function of the form...

IDC IV, ii, fol. 245b, 4-246a, 19 (Latin: 330 E-K):

> We say that the reason why some of the simple bodies move upward, while others move downward, is that each body seeks the place specific to it; for each one of the simple bodies is moved to the place appropriate to it, in the sense that it is its ultimate perfection.... This being so, and because "up" and "down" exist by nature, the mover for these bodies is what generates them as light or heavy, while what is moved is the potentially light or heavy thing, until it exists as light or heavy in perfect actuality: namely, when it exists in its ultimate perfection. But the heavy or light body exists in its ultimate perfection when it exists actually in the place specific to it, namely, "up" [or "down"]. If all this is as we have recounted it, it is evident that the place is the ultimate perfection for these bodies: i.e., the place of each body is the place of the ultimate form of what is generated, because it is the end of the motion which brings it forth from potentiality to perfection.... "Up" is like form for the light thing, and lightness is like matter for it. Similarly, "middle" is like form for the heavy thing and heaviness is like matter for it...

See also *LP* VIII, Comm. 32, 370 FG; *LDC* III, Comm. 28, 198 G; *ibid*., IV, Comm. 22, 249 A-H; *ibid*., Comm. 25, 253 B-E.

95. This condition is stated five times in *Physics* VIII, 4 (255b, 4, 7, 10-22, 23, 24), and frequently in Averroes: e.g., *LDC* III, Comm. 28, 198 G; *ibid*., IV, Comm. 22, 249 C; *LP* VIII, Comm. 32, 370 FG.

96. I have emended the text on the basis of parallel texts and as required by the sense. The *locus* in Aristotle is *Physics* VIII, 4, 255b, 24-27 (note that Simplicius reads *ho to hyphistamenon kai kôlyon aphairôn* at 255b, 24 [*in Physicorum*, p. 1217, 18-19]); cf. also *De Caelo*

IV, 3, 310b, 9-11. In *EP*, the phrase *al-muzayyil al-'â'iq* is used; in *IP*, the phrase is *ha-mêsîr la-makrîaḥ*; in *LP*, it is *auferens impedimentum* (or, *quod aufert impedimentum*).

IP VIII, iv, 4, 2, fol. 121b, 12-14:

> But because these bodies are of this sort, they only make the transition from potentiality to actuality when that which forcibly impedes them is taken away and removed. Consequently, that which removes the impediment must necessarily be their mover, but it is a mover by accident.

LP VIII, Comm. 32, 371 H:

> ...et ita dicitur illud, quod aufert impedimentum, movere lapidem per accidens: et quod generat ipsum, movere ipsum essentialiter.

LP VIII, Comm. 34, 373 B:

> Et similiter diminuit destructionem alterius partis, et est, qui illud, quod movetur ab extrinseco, impossibile est ut moveatur a motore, qui non movetur, nisi per accidens, sicut fuit dictum in auferente impedimentum, et generante in rebus naturalibus.

See also *LP* VIII, Comm. 17, 353 G; *LDC* IV, Comm. 22, 249 C; *ibid.*, Comm. 25, 253 H and 254 A.

All the passages from Averroes hitherto cited describe "that which removes the impediment" as an accidental mover, just as Aristotle does. In *EP* VIII (p. 139, 7-9 Arabic), however, "that which removes the impediment" is called a potential mover:

> Inasmuch as it was demonstrated to his satisfaction that the mover in potentiality is that which removes the impediment, it was also demonstrated to his satisfaction that the elements are not numbered among the things in motion which are moved by themselves essentially.

97. See *Physics* VIII, 4, 254b, 33-255a, 10.

EP VIII (p. 138, 7-9 Arabic):

> Then how is that evident about them? In this way: because, if these bodies were moved by themselves, that is, so that they were moved by themselves without an external mover, they ought to come to rest of the themselves, as is the case [with animals].

LP V, Comm. 18, 221 BC:

> ...quoniam corpora quatuor simplicia existimantur movere se in loco, sed non habent quiescere ex se...secundum hoc igitur animata sunt illa, quae quiescunt ex se.

LP VIII, Comm. 44, 385 C:

> Et dixit: Est illud, quod movet se, et facit ipsum quiescere. quia iam accepit in descriptione moti ex se, quòd est illud, quod habet quiescere ex se. et in hoc differt motum ex se secundum suam opinionem, Å motibus corporum simplicium, cum ista existimantur moveri per se.

98. Narboni (*ad loc.*) says this means something in motion of itself of the genus of the heavy and light. Narboni probably has taken this explanation from *EP* VIII (p. 138, 1-4 Arabic):

> It was already demonstrated in *Physics* VII that those things whose mover is external necessarily terminate at something moved by itself, i.e., like the animals or simple bodies. For that reason, it remained for him to investigate here into what class of moved things the simple bodies fall.

This paragraph of our text begins with a conditional clause, but no proper result clause follows; it is possible that a verb has dropped out at the beginning of paragraph 26. Even so, the meaning of these paragraphs is clear. Averroes is pointing out the limitations of the arguments of *Physics* VII, 1: first, they are based on the self-motion of the elements, but this self-motion is only apparent; second, they are constructed around the impossibility of a simultaneous infinite, without taking into account the impossibility of a sucessive infinite where the members are related essentially. This leads into Averroes' brief discussion of *Physics* VIII, 5, 256a, 4-257a, 33.

99. *Physics* VII, 1, 242a, 15-243a, 2

100. Cf. *Question* VI, the beginning of paragraph 6. **Last** refers here to the extreme member of the series. This may also account for the reading of the Arabic MS in paragraph 26, below: *al-'akhîr*.

101. I.e., those which succeed one another infinitely.

102. Although the text is difficult, the meaning is clear. My emendation here has no support in the MSS, but appears nonetheless to be required.

103. Another emendation may be called for here; perhaps one should read "thing in motion" instead of **mover**.

104. See above, paragraphs 15-16, and the notes *ad loc.* All that is being ruled out here is an infinite series of moved movers in which the cause of motion is always external.

105. Here, animate beings; see paragraph 28.

106. On locomotion as the only form of change applicable to the heavenly sphere, see below, *Question* IX, paragraphs 5-6.

107. *IP* VIII, iv, 4, 6, fol. 126b, 23-127a, 6:

> But inasmuch as we find them [animate beings] set in motion after having been at rest, and at rest after being in motion, so that their proximate mover is moved accidentally, it could be thought that this is possible with regard to the Prime Mover of the universe: namely, that it could be moved accidentally, and that that which is in motion of itself primarily could come rest, as occurs in the case of these animate beings. However, if we give the matter a little thought, it appears that the coming to rest after motion of these animate beings, and their motion after having been at rest, is only a consequence of changes originated in them by that which surrounds them externally, or from themselves.

Narboni explains the words, **not subject to change either because of [a motion produced in] its magnitude**, in our text, to mean "not generated and not corrupted"; while the words, **or because of [a motion produced in] its soul**, mean "not moved accidentally [in place]."

Despite Averroes' description of the first thing in motion of itself as **composed of a mover and something in motion**, the succeeding paragraph of our text makes it very clear that there is no composition involved: the mover does not inhere in what is moved.

108. I.e., the cause of any change produced in these magnitudes.

109. The long lacuna which follows in the Hebrew MSS (up to the words **either essentially or accidentally**) is presumably remedied in the margin of MS. New York, JTSA 2311. Unfortunately, what is written there is illegible. Narboni's sources also appear to have lacked these lines.

110. *KMH*, *Heqqêš*, p. 23a, 11-12 (Latin: *Epitome in Libros Priorum Analiticorum*, 50 L): "An example of this is Hippocrates' saying, 'If man were from one thing, he would not suffer pain.'" Cf. *TAT* II, p. 83 (Arabic: p. 140, 10-15; Latin: 43 D), and for the identification of this quotation, see van den Bergh (*ad loc.*), n. 83.1 (v.2, p. 61).

111. Cf. *LP* VIII, Comm. 45, 385 G-L, especially:

> ...cum declaratum sit quod omnia moventia, quae moventur, reducuntur ad primum motum ex se, quod componitur ex primo moto, et ex motore, quod non movetur, et iam declaratum est quod in nulla hora defecit motus, necesse est ut primus motor, ad quem reducuntur alii motus, sit aeternus, et sit absolutus ab omni transmutatione, et ut motum ab eo etiam sit absolutum ab omni transmutatione, nisi à motu in loco.... Et fuit necesse ut ipse non quiescat à movere, nec illud, quod movetur ab eo à motu, quia est causa motuum.... Et fuit necesse etiam ut non sit transmutabile, neque essentialiter, neque accidentaliter. quoniam, si ita esset, tunc esset motor prior illo...

112. *Physics* VIII, 4

113. *Ibid.*, VIII, 5, 257a, 33 ff.

114. In our text, Averroes presents these words as though they were a direct quotation from Aristotle. These exact words, however, are not to be found in the Greek text of *Physics* VIII, 5, 257a, 26-33, nor in the Arabic version (see p. 852, especially lines 10-12). That this is actually a paraphrase, or comment on these lines from *Physics*, is clear from *LP*, *ad loc.* (Comm. 39, 379 EF), where the words in question do not appear in the lemmata, but the commentary reads:

> ... et si necessarium est in hac perscrutatione consyderare quae istarum rerum motarum est dignior ut ponatur primum, utrum illud, quod movetur ex se, aut illud, quod movetur ex alio extrinseco, quilibet enim homo convenit nobiscum in hoc, quod illud, quod movetur ex se est dignius ut sit primum principium quam illud, quod movetur ex alio.

Cf. *IP* VII, i (quoted above, *Question* I, n. 5), and n. 120, below. Cf. Simplicius, *in Physicorum*, pp. 1040, 30-1041, 5.
On **primarily**, see above, *Question* I, n. 6.

115. Apparently another reference to *Physics* VIII, 5, 257a, 33 ff.

116. Paragraph 30 sets forth the arguments of the last half of *Physics* VIII, 5 in highly abbreviated form. In paragraphs 31-33, Averroes presents the same arguments more systematically, and in more detail. One can only speculate as to the reason for this repetition, but it is found in all our MSS, and evidently existed also in the text available to Narboni. See above, n. 1.

117. Cf *LDC* IV, Comm. 22, 249 FG.

118. **He demonstrated by a general discourse**: I take *yebâ'êr be-ma'amâr kôllêl* as the equivalent of *bî'ûr kôllêl* (cf. in paragraph 33 below, *môpêt kôllêl*). See Wolfson, *Crescas*, p. 328. Averroes refers to this type of proof, e.g., in *LDC* I, Comm. 70, 48 K; *LP* IV, Comm. 87, 173 H-K; *ibid.*, VIII, Comm. 70, 414 G-K (where such proofs are also called *sermones dialectici*). Cf. Simplicius, *in De Caelo*, p. 238, 5-15.

119. *Physics* VIII, 5, 257a, 33-275b, 13

120. **It would more properly be [described as] in motion primarily**: cf. n. 114, above. Note that here the Hebrew MSS (and Narboni) read: "It would more properly be first (*teḥillâ*) in motion primarily...."
Cf. *LP* II, Comm. 3, 49 B-D, where, in still another discussion of the difference between the arguments of *Physics* VII and those of *Physics* VIII, we read:

> ... et causa in hoc est haec, quod Plato opinabatur, quod hic est aliquid, quod movet se. et manifestum est quod, *si hic esset aliquid, quod moveret se, quod illud esset dignius habere hoc nomen natura* secundum hanc descriptionem, scilicet secundum quod

> dicitur hic, quod naturalia sunt illa, quae non moventur ex extrinseco. et manifestum est quod quicquid non movetur ex extrinseco, movetur ex se.

On what follows in our text, see *Physics* VIII, 5, 257b, 13 ff.

121. *Physics* VI, 4

122. I.e., the series of moved movers must proceed in only one direction, or return upon itself.

123. *IP* VII, i, fol. 102a, 3-5:

> ... for if nothing in motion primarily exists in that which is in motion of itself, an actually infinite number of parts would exist in that which is in motion: because it would be necessary for all of them to cause motion and undergo motion simultaneously.

124. **It will not be moved primarily**: while this statement is true, what we would expect to find here is, "and there will be no part that is moved primarily."

125. I.e., there will be nothing actually in motion. Narboni comments:

> As for the infinite rectilinear series which Averroes mentions here, it is not impossible in the same way as other kinds of self-movers... and for that reason, Aristotle treated this by itself, that is, not confused with the investigation of that which moves itself, so that he could demonstrate to his satisfaction that there is something moved by itself primarily, and so that there should be nothing to dispute with regard to the fact that not every mover is moved, even if there is nothing that moves itself.

126. *EP* VIII, pp. 142, 8-143, 8 (Arabic):

> We say that if something moves itself, either all of it moves and is moved simultaneously in one respect... or one part moves another part and moves itself. Whichever of these alternatives we assume, it follows that the thing moves and is moved simultaneously in one respect, which is absurd. For the mover is a mover in so far as it is actual, and what is moved is moved in so far as it is potential, as has been demonstrated from the definition of motion.... Now, if we assume that part of it moves another part, and that other moves a third part, either the causing of motion does not return circularly so that the first part moves and is moved simultaneously; or the causing of motion does return circularly, so that the third part, for example, moves the first. From this it also follows necessarily that the thing moves itself; for there is no difference between

a thing moving itself essentially without an intermediary, or through more than one intermediary.

IP VIII, iv, 4, 5, fol. 124a, 10-124b, 8:

> We say that inasmuch as it has already been demonstrated that everything in motion is divisible and that everything divisible is a body, it is possible that in that which is in motion of itself, because it is a body, either the whole moves itself and is moved by itself simultaneously... or that each part of it moves every other part.... If the whole moves itself and is moved simultaneously, it follows that that which causes motion in place undergoes motion in place... and, in general, that that which is the thing actually is itself the same thing potentially.... If we assume that each part of it moves every other part, that is, that they move one another circularly, it follows that all of them move themselves and are moved by one another; for each one of them is moved by something other than itself accidentally, and moves itself by the mediation of something other than itself.

127. See above, n. 118.

128. I.e., because their apparent self-motion is a consequence of motions produced in them by other movers. See above, paragraph 28, and *Physics* VIII, 6, 259b, 1-20.

129. See *LP* VIII, Comm. 52, 392 F-H.

130. See *Physics* VIII, 6, 259b, 20-28. Note, however, that there the argument is couched in terms of the Prime Mover, while in our text the emphasis is on that which is moved of itself essentially. In the Arabic version of *Physics* (p. 872, 4 ff.), the argument is put in terms of *motion*.

131. See *Physics* VIII, 6, 259a, 6-13 where, again, the argument is concerned with movers, as in *LP* VIII, Comm. 48, 389 A-D. Cf. *LP* I, Comm. 50, 31 EF, where the argument deals with the principles of generation and their number.
IP VIII, iv, 4, 6, fol. 126a, 15-23:

> And this is one of the arguments from which it is evident that there is an eternal mover and something in motion of itself which is eternal, whether one or more than one; [and in the latter case,] whether finite in number or infinite in number. But it is preferable, as Aristotle said, that we should assume it to be one, not many, for it is sufficient to assume it to be one. If, however, they are many, it is preferable that they should be finite in number; for if one and the same thing, or many identical things, may exist from a finite number of things just as well as from an infinite number of things, it is preferable that

the things from which they exist should be finite in number. But such an investigation is not proper to this place.

132. **Those around which the things perpetually in motion are moved:** the Hebrew phrase is ambiguous (*we-hêm 'asher ʻalêhem yitnôʻaʻû ha-mitnôʻaʻîm tamîd*). In this context, it could be taken to mean the Prime Mover which is perpetually at rest and which supports the universe (as in *De Motu Animalium*, 1-4; cf. *Metaphysics* XII, 7; and see W. Jaeger, *Aristotle* [Oxford, 1934], ch. XIV). It could, however, refer to the earth, over which (or around which, as center) the heavens move (see *De Caelo* II, 3, 286a, 13-21 and 286b, 8-9; cf. *Physics* IV, 4, 212a, 21-14).

We have a large number of passages in which Averroes expresses the latter view: e.g., *EP* IV, p. 15b, 22-26 (Arabic: p. 54, 4-7):

> The proof that what is moved circularly requires a spherical body at rest around which it is moved [and] which is not a part of it, but is separated from it and meets it through contact, is that the natural sphere, *qua* sphere, cannot possibly be without a center around which it moves [and] which is the cause of its being stable in its totality and in motion with respect to its parts.

Ibid, IV, p. 16a, 16-20 (Arabic: p. 55, 3-5):

> Furthermore, it is one of the things known immediately that that which is in motion is not moved around itself, and that everything in motion requires something at rest around which it is moved. For this reason, it follows necessarily that the center is separate and at rest, and what is of this nature is necessarily a body.

See also *LP* IV, Comm. 43, 142 L:

> Ad hoc autem dicendum est qui illa, quae moventur per se, indigent aliquo quiescente, circa quod moventur, ut declarat Ari. in lib. de Motibus animalium localibus: et hoc quiescens forte erit locus per se, quando non fuerit continens rem motam, et forte erit locus per accidens, quando non fuerit continens rem motam in omnibus partibus, sicut est dispositio in corporibus coelestibus ...

See *ibid.*, VIII, Comm. 19, 354 I; *ibid.*, Comm. 84, 432 F.

EDC I, pp. 28, 18-29, 2:

> Furthermore, it has already been demonstrated in the first part of this book that in its very nature this body is neither generated nor corrupted because it has no contrary. If this is so, and it is from the necessity of existence of this circular body that the rest of the elements exist--for there is no doubt that there must be something at rest over which it is moved circularly, as has been proved--it follows necessarily that the world in its totality must be eternal.

Ibid., II, p. 42, 16-19:

> ... for inasmuch as there is an eternal mover, there must necessarily be an eternal thing in motion; and inasmuch as there is an eternal thing in motion, it follows necessarily that there must be a body at rest over which it is moved circularly, namely, the earth.

IDC II, ii, 2, fol. 184b, 11-13 (Latin: 297 K):

> If there is a circular body moved with a perpetual motion, then necessarily there must be something fixed over which it is moved. For it is impossible for anything to be moved without something at rest over which it is moved; and the thing of this description is the earth.

IP VIII, iv, 4, 6, fol. 128a, 1-13:

> From this, that which was the subject of our investigation from the beginning is indeed clear: namely, that some of the natural things which are in motion are in motion eternally, and some of the natural things are at rest eternally, that is, those things around which the things that are in motion eternally are moved.

See also *LDC* I, Comm. 97, 65 K; *ibid.*, II, Comm. 102. Cf. Themistius, *in De Caelo*, p. 66. 10-11; the interesting passage in Simplicius, *in De Caelo*, pp. 403, 7 ff.; and *TAT* XIV, p. 291 (Arabic: p. 478, 3 ff.; Latin: 118 I), reading with Bouyges rather than van den Bergh. See also Wolfson, *Crescas*, pp. 451-54; and Hugonnard Roche, "L'Épitomé," pp. 35-37.

There are also passages in which the referent is unclear (although it is likely that they, too, refer to a physical entity): e.g., *EP* VIII, p. 137, 4-5 (Arabic):

> And if it has been demonstrated that things like these are eternal with respect to motion, then it follows necessarily that there must also be something which is eternal with respect to rest, around which this motion takes place.

Finally, we have the statement of *LP* VIII, Comm. 21, 356 H-K, where Averroes explicitly recognizes and discusses the ambiguity of the phrase, suggesting that "what is eternally at rest" may refer to the Prime Mover:

> Modis autem declarationis istius intentionis de praedictis talis est. Declaratum est quidem in Septimo ab Arist. hoc esse primum motum in loco ex se, ad quem perveniunt caeteri motus locales. et est manifestum per se quod caeteri motus inveniuntur per sensum reduci ad motum in loco, scilicet ad corpus coeleste. Et, cum hoc fuerit coniunctum ad illud, quod declaratum fuit primo in hoc tractatu, scilicet quod primum motum est aeterni motus, erit necesse hic esse aliquod quiescens aeternum, super

> quod moveatur hoc primum motum. et hoc motum aeternum, quod declaratum est, est illud, quod videtur sensu moveri semper: et est corpus coeleste, et quiescens, quod videtur sensu semper quiescere, est terra. et est illud, quod declaratum est ratione. nihil enim est hic, quod posset convenire ei, quod declaratum est ratione, nisi istud. Et forte illud, quod videtur semper quiescere, est primus motor istius motus, et est illud, quod post declarabitur. Sed istud non dicitur quiescens, nisi aequivoce.

133. **Each one of the things:** the Hebrew reads, "each one of the two of them." This indicates that the translator, or the text from which he worked, read *minhumâ* rather than *minhâ*.

134. See *De Gen. et Corr.* II, 10, 336b, 3-9.
IP VIII, iv, 4, 6, fol. 128a, 13-17:

> ... and inasmuch as some things are moved at one time and are at rest at another, for those things which are moved by an eternal unmoved mover must be moved eternally, while those which are moved by an eternal moved mover must be moved at one time and at rest at another, in accordance with whether that [mover which is also] in motion is close to them or remote from them...

See also *LP* VIII, Comm. 9, 345 FG; *ibid.*, Comm. 21, 356 K; *ibid.*, Comm. 46, 386 M-387 A; *ibid.*, Comm. 53, 394 D; *IDGC* II, p. 89, 29-41 (Latin: pp. 147, 12-148, 9 Fobes); *EDGC* II, p. 123, 90-96 (Latin: 395 [col. 4]; Arabic: p. 29, 8 ff.).

135. *Physics* VIII, 5, 285a, 18-20

136. The reading of the Arabic MS is correct: if it is moved accidentally, it cannot cause eternal motion; if it is moved essentially, it can cause motion in other things only when it stands in the proper relation to those things.

137. See *LP* VIII, Comm. 53, 393 L-394 B, and *ibid.*, Comm. 50, 390 E.
For were this the case, the mover would be subject to motion accidentally, and [that in which it causes motion] would not be the first of things in motion of themselves: Narboni comments:

> [These words mean] that if the mover were a force in a body, and mingled with what is moved by it and distributed throughout its dimensions, the mover would itself be moved by accident. Similarly, what is moved would not be moved by itself in its totality, except accidentally, for it would be moved because of a mover itself moved accidentally; and it would not be the first of things moved by themselves: that is, it would not be moved primarily by the first mover, but rather by the mediation of a mover other than itself, inasmuch as the true Prime Mover is not moved, for there must be some first thing in motion primarily of itself for whose motion something else in motion is not a cause.

> This is the celestial body which contains the universe, for it is truly moved by itself primarily: it is not accidentally moved by something external, nor by a mover itself moved accidentally. In truth, it is compounded of only two parts: a motive part which is not itself moved either essentially or accidentally, because it is separate from the body in so far as it is not mingled with it, while it is in it in so far as it bestows motion and existence; and a part which does not cause motion to any other part, for it is deprived of divisibility and finitude and corruption because it is the dwelling of God. For this reason, it is truly what is moved primarily, as the One, true God is the Prime Mover.

Cf. *DSO* VI, p. 135 (Latin: 13 CD; Hebrew: p. 57, 127-134).

138. Cf. Simplicius, *in De Caelo*, p. 403, 15-21.

139. See *LP* VIII, Comm. 53, 393 L-394 B; cf. *ibid.*, Comm. 50, 390 E. On the types of demonstration involved, see *Post. Anal.* I, 13, and *KMH*, *Môpêt*, p. 35b, 18-30 (Latin: *De Demonstratione*, 53 E-G):

> ... the objects of research which we desire to establish by means of the demonstrative syllogism are of three kinds: either we seek to establish the cause of the thing and its existence simultaneously; ... or we seek to establish its cause alone, when the knowledge of its existence has reached us through [immediate] knowledge or syllogism; or we seek to establish its existence alone.... The first kind informs through absolute demonstration, namely, demonstration of the cause and existence simultaneously; the second informs through *demonstratio causae*; and the third informs through *demonstratio essendi* and *demonstratio per signum*.

Ibid., p. 40b, 10-30 (Latin: 56 K-M):

> As for the *demonstratio causae* alone, its parts are arranged [as are] the parts of the simultaneous *demonstratio essendi et causae*, and [its] conditions are these same conditions.... But the difference between them is that in this proof the existence is known to us, either through immediate knowledge or through syllogism, so that through this proof we seek only to establish the cause.... The conditions of the *demonstratio essendi* are these same preceding conditions, except that the middle term in it is not a cause for the conclusion; rather it is a cause for our knowledge alone, and what is prior in it in knowledge is not the same as what is prior in existence. What is of this description is of two kinds: one kind, and it is the first, is what is called *demonstratio per signum*. This is the kind in which the prior in existence is always demonstrated by what is posterior to it, if the latter is prior in knowledge. For example, we demonstrate that the moon has a spherical shape because its light increases in the lunar shape...

Cf. *QLog*, *libri posteriorum*, VI, 114 M-116 A, and *IPsA* I, pp. 71-73. See also Narboni *ad Moreh Nebukim* I, 71:

> ... and the eternity of motion proves the eternity of the mover by way of *demonstratio essendi*, as the eternity of the mover proves the eternity of what is moved and of the motion by way of *demonstratio causae et essendi*.

In his commentary on this *Question*, Narboni notes that the *demonstratio causae* applies to the motion, not the mover, for *demonstratio causae* with respect to the latter is impossible. Cf. Narboni *ad Moreh Nebukim* I, 71:

> ... he means by "decisive demonstration" what Aristotle called "absolute demonstration," namely, *demonstratio causae et essendi*. And this is true, for this kind of demonstration, whose existence is very rare, does not often exist with respect to physical things. The existence of God is also proved by *demonstratio per signum*, namely, from the posterior, not from the prior; for He is prior to all things, and there is nothing prior to Him, and how could His existence be proved by a *demonstratio causae* when He is the cause of everything?

Cf. *LP* II, Comm. 22, 57 AB; Alexander, *De Prin*., p. 253.

140. On the illustration, see *De Caelo* II, 11, 291b, 17 ff.; *Posterior Analytics* I, 13, 78b, 4-10.

141. See above, paragraph 20.

142. *Physics* VIII, 7

143. The reference is probably to *Physics* VI, 10, 241b, 2-20; but cf. *ibid*., VIII, 8, 261b, 31-263a, 4, and 264a, 28-33; *ibid*., VIII, 9, 265a, 13-32. See also *LP* VI, Comm. 91.

144. **Contiguous**: in Hebrew, *kerûkâ* = Greek *echomenon*; see Wolfson, *Crescas*, pp. 375-76, n. 113. Contiguity is only a special case of succession (*Physics* V, 3, 227a, 4-22; *Metaphysics* XI, 1069a, 1-13), and here the Hebrew terms *kerûkâ* and *mesudderet* may be used synonymously (note that at *Physics* VIII, 10, 267a, 24 and 267b, 15, Aristotle uses *echomenon* in the sense of *ephexês*). At *LP* VIII, Comm. 56, 397 F (to *Physics* VIII, 7, 260b, 21 ff., where Aristotle uses *ephexês*), Averroes comments:

> ... aut quia componitur ex motibus successivis, scilicet quorum unus est ante alium, et unus post alium, ita, quod finis praecedentis sit cum principio consequentis.

(Cf. Gersonides, *Milḥâmôt ha-Šêm* VI, i, 11, p. 341, 26: "... for there cannot be an eternal, continuous time from the successive motions.")

145. *Physics* VIII, 10, 267a, 21-24, and 267b, 9-11; cf. *LP* VIII, Comm. 55, 396 EF.

146. *Physics* VIII, 8, 261b, 27-262a, 15; 264a, 28-264b, 1; 264b, 13-24; cf. *LP* VIII, Comm. 64, 404 A, E.
EP VIII, pp. 149, 17-150, 10 (Arabic):

> With respect to this investigation, it remains for us to demonstrate which motion is the one that has such a mover. We say that it is clear from what was demonstrated that it is the motion of translation, for this is prior by nature to the rest of the motions... [and] also because that which is moved with this motion is moved by itself, and such a thing can exist only with reference to [motion in] place, and of that [kind of motion], only in the circular. For it is not possible for any motion which is one, continuous, and eternal to exist except this [motion], inasmuch as the rest of the motions, according to what was demonstrated in the sixth book, are finite. Furthermore, it is not possible for the motion which is compounded of two of these rectilinear motions to be one and continuous...

147. See *Physics* VIII, 6, 259a, 17-20; cf. *LP* VIII, Comm. 83, 431 L-432 A; *LM* XII, Comm. 43, 326 LM (Arabic: pp. 1644, 7-1645, 1); Alexander, *De Prin.*, p. 267.
IP VIII, iv, 4, 6, fol. 126a, 23-126b, 8:

> It is already evident, also, from two aspects, that there must be an eternal mover, for it has already been demonstrated that there must be a motion which is one and which does not end. If there must be a motion of this kind, it must also be continuous. If it is continuous, it must be one: for the contiguous and successive motions are neither continuous nor one, but many. That the eternal motion must be continuous is evident because the motions which are contiguous cannot be eternal in themselves, for they are infinite [in number] and from an infinite [number] of movers. Thus, if there must be a motion which is continuous and one, there must be a mover which is one and eternal, and there must be something moved by this mover which is one and eternal...

Ibid., VIII, vi, 5, fol. 139a, 16-22:

> When it has been demonstrated that there is a continuous motion, and that the one, continuous motion belongs only to one thing in motion and one mover, and that otherwise the motions are contiguous and not one, continuous motion: it is evident that what causes that motion and what is moved by it are each one thing, and that this mover cannot be moved; for the series must proceed to an unmoved mover, inasmuch as everything in motion has a mover but this cannot continue to infinity...

IDC II, v, 2, fol. 190b, 9-13 (Latin: 301 CD):

> The second proposition is what was demonstrated about the prime mover and the first thing in motion (i.e., the celestial body and its mover): neither one of them is capable of change. If either of them were capable of change, the primary motion would not be continuous [and] eternal, as was demonstrated in *Physics* VIII.

EM IV, 385 M (Arabic: p. 129, 7-8; German: p. 107): "Et si datur motus aeternus, necessario quoque datur motor aeternus unus. nam, si essent multi, non esset unus motus continuus."

148. I.e., by considering each of the four varieties of motion; see *Physics* VIII, 7, 260a, 20-261a, 26. Cf. Wolfson, *op. cit.*, p. 628, n. 2.

149. *Physics* VIII, 7, 260b, 18-19 (cf. 261a, 15-21). On **perfection of existence**, see above, *Question* II, n. 10.

150. **[In] the world as a whole, [not] in the individual**: the Hebrew reads: *û-be-'îsh ha-'ôlâm bi-klâlô*. These words baffle me. I have translated as if this were a reference to *Physics* VIII, 7, 261a, 13 ff. (as in *LP* VIII, Comm. 1, 339 EF; *ibid.*, Comm. 57, 398 D-H; cf. *Physics* VIII, 2, 252b, 24-28): "In the individual, locomotion is the last of the motions; whereas it is the first, and only, motion of the heavens." See Cleary, *Priority*, pp. 80-85.

Possibly, however, these words reflect something similar to what is found in Simplicius, *in Physicorum*, p. 1272, 32-33: that the self-moved is most peculiarly and properly moved with motion in place (*kai hoti tên kata phoran kinêsin idiôs kai kyriôs kineitai to autokinêton*).

151. *Physics* VIII, 5; cf. *ibid.*, VIII, 10, 267a, 21-267b, 5.

152. Clearly, either the bracketed words must be added to the text, or we must emend **is moved** to "he causes."

153. See *Physics* VIII, 10. Cf. *LP* VIII, Comm. 78, 424 A, and the treatment of the same arguments in *Question* VIII, below.

154. *Physics* VIII, 10, 266a, 12-23; cf. *Metaphysics* XII, 7, 1073a, 3-8. Averroes apparently agrees with Simplicius that the "nothing finite" of *Physics* VIII, 10, 266a, 12-13 refers to force rather than magnitude (see Ross' commentary to these lines in his edition of *Physics*, p. 721), but perhaps in this very brief recapitulation Averroes simply prefers not to consider the case of magnitude. Our text (and below, *Question* VIII, paragraphs 3-4) agrees with *IP* VIII, vi, 1, fol. 137a, 8-9:

> It would appear that he employs the *demonstratio per signum*, for it is one of the things known immediately that what causes motion for an infinite time has within it an infinite force.

(But note that in *Question* VIII (*loc. cit.*), this proposition is demonstrated.)
However, in *LP* VIII, Comm. 79, 425 H, we read:

> ... et hoc, quod dicimus motor finitus, intelligimus motorem finitum potentia, et magnitudine. sed hoc quidem, quod motor, cum fuerit corpus, sit finitae magnitudinis, iam probatum est in Tertio: quod autem finitae potentiae, cum sit corpus finitum, hoc nondum manifestum est...

155. *EP* VIII, p. 147, 12-13 (Arabic):

> That it is not possible for a corporeal force to have an infinite action is clear in the following manner: because every corporeal force must necessarily be divided when the body is divided.

See also *LP* VIII, Comm. 78, 424 B; *LM* XII, Comm. 41, 324 A (Arabic: p. 1627, 1-4).

156. The impossibility of an infinite body is demonstrated in *Physics* III, 5, and *De Caelo* I, 5-7. The argument given in our text against the existence of an infinite force in a body is not found in *Physics* VIII, 10. Rather, the argument of our text seems to be based on an analogy with the arguments of *De Caelo* I, 6 (against infinite weight; cf. *ibid.*, I, 7, 275a, 1 ff.).

157. Cf. *Metaphysics* XII, 7, 1073a, 3-11; and *IP* VIII, vi, 2, fol. 138a, 4-10:

> After it had already been established that it is necessary for the eternal motive force to be infinite, and that it is impossible for an infinite force to be a body or in a body, it is evident that the eternal Prime Mover is neither a body nor a force in a body; and that it is not divisible or subject to change accidentally, in the way that the souls of animals undergo motion through the motion of their bodies, because these souls have no existence save through their bodies. But this mover must necessarily be incorporeal, and its existence is not through bodies; rather, the existence of the [celestial] body is through it.

Ibid., VIII, vi, 2, fol. 137a, 10-20:

> When it has been settled that this motion [whose existence was] demonstrated in this book is infinite and cannot exist as the result of a mover of finite force, it must necessarily exist as the result of a mover of infinite force. I say that this force cannot exist within a body or throughout a body. The proof of this is that every body is either finite or infinite, but it has already been demonstrated that the existence of an infinite body is impossible. This motive force, therefore, must exist

> throughout a finite body or within a finite body. But the existence of an infinite force throughout a finite body or within a finite body is obviously impossible, for it is evident that bodies or forces which exist in bodies are divided when the bodies are divided.

EP VIII, p. 148, 11 [reading with the Cairo and Madrid MSS]-12 (Arabic):

> In this way it is clear... that a finite force cannot have infinite action. It has already been demonstrated, therefore, that the motive force which belongs to a body such as this is not material.

See also *LP* VIII, Comm. 79, 425 H-426 G; *EM* IV, 385 M-386 A (Arabic: p. 129, 8-17; German: pp. 107-108); *LM* XII, Comm. 41, 323 L-324 B (Arabic: pp. 1626, 6-1628, 9); *DSO* VI, pp. 126-27 (Latin: 11 K-M; Hebrew: p. 53, 28-54, 43 ; *ibid.*, III, pp. 99-100 (Latin: 8 I-L; Hebrew: pp. 36, 1-37, 15).

158. *KMH*, *Heqqêš*, p. 22b, 2-11 (Latin: *in Libros Priorum Analiticorum*, 50 FG):

> The *demonstratio per impossibile* is a compound of categorical and hypothetical [proof], and it is employed in the following way: when we wish to demonstrate the truth of a certain judgment, we take its contradictory and to this we join a correct proposition about whose truth there is no doubt. From this, we form one of the valid combinations in whatever figure of the categorical syllogism that is convenient. When a demonstrably false conclusion is deduced, we know that the falsehood comes neither from the combination of the syllogism (for it was a valid combination), nor from the correct proposition. Thus, the only remaining alternative is that the falsehood comes from the contradictory of the proposition in question, and inasmuch as its contradictory is false, the proposition itself must be true.

Cf. Maimonides, *Millôt ha-Higgâyôn* VII, 5, pp. 39-40 (Arabic: pp. 15, 19-16, 5 [Efros]):

> The men of this art have a type of syllogism which they call *demonstratio per impossibile*--for when we wish to verify a certain judgment and we proceed by means of some categorical syllogism amd deduce the judgment whose truth we wish to know, we call it the direct categorical syllogism. If we verify that judgment in another way--that is, we assume the contradictory of the judgment whose truth we wish to know, and we compose a syllogism through which we have a demonstration, and the falsehood of that contradictory which we assumed is deduced: then the contradictory of what we assumed is undoubtedly the truth, namely, the judgment which we wished to verify. We call this syllogism, by which the falsehood of the contradictory of the

judgment we wish to verify was demonstrated, *demonstratio per impossibile.*

See also *Prior Analytics* I, 23, 41a, 21 ff.; *ibid.*, I, 29, 45a, 23 ff.; *ibid.*, II, 11; *IPrA* II, cap. xi, 132 H. Note especially *LDC* I, Comm. 95, 64 D: ... *demonstrationes autem ducentes ad impossibile sunt generis signorum.*

159. See *Physics* VIII, 10, 266a, 24-266b, 5. The only impossibility mentioned by Aristotle at the end of the book is the impossibility of an infinite body (267b, 20-22). Cf. *LP* VIII, Comm. 79, 426 GH.

160. In his various formulations of these alternatives, Averroes seems to use the terms "force" and "motion" interchangeably. See, e.g., *DSO* VI, pp. 134-35 (Latin: 13 BC; Hebrew: p. 57, 119-126), and below, n. 162.

161. Following (more or less) Aristotle's argument, Averroes does not distinguish here between infinite in intensity and infinite in duration. Here we have his interpretation of *Physics* VIII, 10, 266a, 28-31, as in *EP* VIII, p. 148, 8-11 (Arabic):

> The motion of this body from the infinite force must take place in an instant, and this is an impossible self-contradiction: for it has already been demonstrated that every motion takes place in time. It is obvious that this absurdity follows only from our assumption of an infinite force in a finite body.

See also *LP* VIII, Comm. 79, 425 M; *DSO* III, pp. 100-102 (Latin: 8 L-9 A; Hebrew: p. 37, 16-23).

162. **... the motion of the part is smaller than the motion of the whole...**: all the MSS read: "... the motion of the whole is smaller than the motion of the part...," which cannot be correct. See, e.g., *EP* VIII, p. 148, 15 (Arabic): "... because the force of the part, for example, is smaller than the force of the whole"; and *DSO* VII, 13 IK (Hebrew: MS P. 989, fol. 29a, 14-17):

> ... For it is already evident that the force of a part of [one of] those bodies whose force is divided when they are divided is smaller than the force of the whole. And this proof includes both the active and passive forces which exist in every body, according to what appears in the first book of *De Caelo*. And if this is so, then every force in a body, whether active or passive, is necessarily finite: for the force of the whole in such bodies is greater than the force of the part.

While the discussion of instantaneous motion clearly implies that Averroes has motion infinite in intensity in mind, his discussion of the relation between the motion of the part and the motion of the whole could also apply to motion infinite in duration. Cf. *LP* VIII, Comm. 78. 424 BC, where this argument is given in terms of "force."

163. Again, this is not an argument found in *Physics* VIII, 10. For the full form of the argument, see *LP* VIII, Comm. 78, 424 BC, where the argument occurs in an excursus and does not stem directly from the text of *Physics*.

164. Cf. *LP* VIII, Comm. 86, 433 I-M. In view of what follows in our text, especially in the first line of paragraph 47, below, perhaps **in this [eighth] book** (*be-zeh ha-ma'amâr*) ought to be translated: "in this chapter."

165. Averroes discusses the question of matter and form in the heavens in a variety of contexts. See *LP* VIII, Comm. 79, 426 M-427 B; *ibid.*, Comm. 78. 424 IK; *ibid.*, I, Comm. 63, 38 F; *LDC* I, Comm. 5, 5 EF; *ibid.*, Comm. 20, 15 CD; *ibid.*, Comm. 92; *ibid.*, II, Comm. 1, 95 LM; *ibid.*, Comm. 3, 97 A-C; *ibid.*, Comm. 36, 120 E; *ibid.*, Comm. 71, 146 FG; *LM* VIII, Comm. 4, 211 F (Arabic: p. 1032, 10-14); *ibid.*, Comm. 12, 220 G-I (Arabic: pp. 1077, 19-1078, 16); *ibid.*, XII, Comm. 10; and *DSO*. See also below, *Question* IX, paragraph 5, and the notes thereto; and Wolfson, *Crescas*, pp. 594-95.

The lack of matter and form in the heavens is largely deduced from *De Caelo* II, 3.

166. See above, nn. 144 and 147.

167. This discussion depends on *Physics* VIII, 10, 267a, 21-25. For parallels, see above, nn. 137 and 147.

168. **... because it is exceedingly obvious from what had been demonstrated concerning the [prime] mover...**: Narboni takes these words to mean that this point was dealt with potentially earlier on when Averroes said **and it has already been demonstrated of the Prime Mover of the universe...** (paragraph 28).

169. Cf. *LDC* II, Comm. 36, 120 GH.

170. Cf. paragraphs 31-33, above.

171. Cf. *EDC* II, p. 41, 8-14:

> And this is what aids our understanding of the interpretation of Aristotle's discourse at this point. By the discourse about these things here, I was reminded of what had previously been demonstrated about the heavenly body: that it is animate--and this exists potentially in the preceding arguments, for it was demonstrated that the heavenly body is moved in virtue of itself. And it is also evident that it is animate, as Alexander said, because it is eternal, and the eternal is better than the animate thing which is not eternal, and best animate thing is necessarily animate because it is better.

See also *DSO* VI, pp. 130-31 (Latin: 12 FG; Hebrew: p. 55, 71-78), and cf. *TAT* XIV, pp. 287-288 (Arabic: pp. 472, 7-473, 14; Latin: 117 E-H); *ibid.*, III, p. 114 (Arabic: p. 190, 3-8; Latin: 53 M-54 A).

172. This may refer to Plato and his followers (as in *Question* I; see below, n. 175), or, possibly, to Avicenna and his followers (see below, *Question* VIII, paragraphs 6-7, and n. 43). As we have seen, Averroes does not always clearly distinguish between the opinions of Plato and those of Avicenna. Indeed, the opinions of the Mutakallimûn and John Philoponus are often considered to be identical with those of Plato and Avicenna. Cf. Evrard, *op. cit.*, pp. 307-13.

Here, the term **external** must mean an immaterial mover which, unlike animal and human souls, does not require a body in order to exist. See H. Wolfson, "The Problem of the Souls of the Spheres from the Byzantine Commentaries to Aristotle through the Arabs and St. Thomas to Kepler," *Dumbarton Oaks Papers* 16 (1962), especially pp. 33, and 42-44.

Cf. *DSO* VI, pp. 123-24 (Latin: 11 E-H; Hebrew: pp. 52, 1-53, 19) for a parallel couched in slightly different terms.

173. I.e., bodies with immaterial movers, such as animals. Aristotle and Averroes have already shown that the Prime Mover does not inhere in the celestial body as the soul of an animal inheres in the animal.

174. Cf. *Question* I, paragraph 6 (also paragraph 4, and n. 12), where a very similar phrase is used in connection with Plato's theory of the self-mover. Here, too, Averroes is denying that the heavens can move themselves in such a way that the mover and what is moved are identical. I take the phrase **either potentially or actually** to refer to the distinction between the mover and what is moved: if the self-mover exists, it must be simultaneously potential (what is moved) and actual (the mover), which is impossible.

175. What is simultaneously potential and actual cannot be one.

176. Cf. *Question* IX, the end of paragraph 6. According to Averroes, his own view is opposed to the opinions of Alexander (see *Question* IX, paragraph 8; *DSO* VII, 14 E [Hebrew: MS P.989, fol. 29, 42 ff.]), Plato (see *LP* VIII, Comm. 79, 426 K), and Avicenna (see *Question* IX). See also *IDC* I, 293 GH and 294 C-295 B; *LDC* II, Comm. 71, 145 K.

177. Having disposed of the possibility of an infinite force within a body, Averroes now turns to the case of an external force (or body), infinite with respect to intensity.

178. **... the celestial bodies would destroy all the moved bodies in the sublunar world**: I have not been able to find this striking phrase elsewhere. Possibly it is related to *De Caelo* II, 9, 291a, 21-22, but cf. *DSO* VI, p. 134 (Latin: 13 B; Hebrew: p. 57, 119-20).

179. *LP* VIII, Comm. 79, 427 G: *quanto enim maior fuerit motor, tanto motio eius erit velocior.* Cf. *ibid.*, 427 F.

180. Cf. *LDC* II, Comm. 71, 146 C-E; *ibid.*, Comm. 38, 122 F-K.

181. There would be no relation because the infinite has no relation to the finite. See *De Caelo* I, 6, 274a, 7-8; cf. *IDC* I, 294 B; *LP* VII, Comm. 35, 335 B-E.

182. See *LP* VIII, Comm. 79, 427 BC; *DSO* III, p. 101 (Latin: 8 M; Hebrew: p. 37, 17-18); *LDC* II, Comm. 63, 141 GH; *IDC* I, x, 2, 8, fol. 178b, 3-6 (Latin: 293 M-294 A):

> ... it is indeed evident that they cannot have that kind of lack of finitude which pertains to the intensity of motion and its velocity: for if they were infinite in this respect, it would be possible for there to be motion in an instant.

183. *IDC* I, x, 2, 8, fol. 178b, 6-9 (Latin: 294 A):

> This kind of finitude [with respect to intensity] is necessitated by the nature of what is moved, *qua* body. For this reason, this kind of finitude must include everything moved by itself, whether or not the form through which it is moved inheres in the body.

Cf. *LP* VIII, Comm. 79, 427 C.

184. *EP* VII, p. 39a, 2-3 (Arabic: pp. 125, 15-126, 2):

> ... for anything whatsoever does not move anything whatsoever; rather, only a defined thing moves some [other] defined thing, and with a defined force.

EDC IV, p. 67, 15-17:

> ... and as what was moved there was not moved from anything whatsoever to anything whatsoever... but, rather, from a definite thing to a definite thing; nor, again, is the mover anything whatsoever...

IDC II, v, 4, fol. 192a, 1-3 (Latin: 301 M-302 A):

> ... and this is apparent with regard to every natural mover: it has a defined proportion to what is moved [by it]. For this reason, whatever mover that happens to exist does not move whatever moved thing that happens to exist, nor does it do so in whatever time happens to exist.

Ibid., IV, ii, fol. 245b, 11-16 (Latin: 330 FG):

> With respect to all motions, it is clear that no one of them is from something indefinite to something indefinite; nor is the thing in motion which is described [as being moved] by one of these motions something indefinite; nor is the mover of any one of these motions something indefinite. Rather, it is evident

> that in each one of the classes of motion, [something] is moved from a definite opposite to a definite opposite.

See also *LM* XII, Comm. 41, 324 E-G (Arabic: pp. 1629, 16-1630, 11); *LDC* I, Comm. 64, 44 DE; *ibid.*, IV, Comm. 22, 248 GH. Cf. *TAT* III, pp. 112-114 (Arabic: pp. 187, 13-189, 13; Latin: 53 F-L).

185. **... a mover might cause the same motion in the larger [object] as in the smaller**: the text is ambiguous. In addition to the translation given, it might mean, following the reading of MS. Munich 31, that two movers of different magnitudes could cause a motion of the same magnitude or intensity in the same object. Inasmuch as what is important here is the proportion between *movens* and *motum* (active potentialities amd passive potentialities), neither of the alternatives would appear more correct than the other. See *LM* XII, Comm. 41, 324 E-H (Arabic: pp. 1629, 16-1631, 7); *LDC* II, Comm. 38, 122 G-K; *ibid.*, Comm. 63, 141 F; *LP* VIII, Comm. 35, 335 B-D.

186. On this objection, see Lewin, "*Muḥdath*," p. 91; Pines, "Summary," pp. 322-25; Wolfson, *Kalam*, pp. 374-82. The text of this *Question* (if I have punctuated the text correctly) would appear to support Wolfson's reconstruction of the original form of Philoponus' argument. Cf., however, *LP* VIII, Comm. 79, 426 K-M, where another objection raised by John Philoponus is said to be the strongest, and the present argument is attributed to Avicenna and Alexander (on which, cf. *Question* IX, below).

Simplicius reports the argument (*in De Caelo*, p. 79,1 ff. [tr. Wildberg, p. 66]), but lays no special stress on it.

187. The first part of this paragraph (through **which may be raised in this regard**) is translated in Steinschneider, "Alfarabi," p. 122. On the argument, cf. *Question* IX, below.

188. *De Caelo* I, 12; see also the almost verbatim parallel to our text in *LM* XII, Comm. 41, 324 BC (Arabic: p. 1628, 10-15):

> John the Grammarian raised a great and difficult question against the Peripatetics. He said that every body has a finite potentiality, and the heavens are a body. Therefore, the heavens have a finite potentiality. But everything finite is corruptible, so that the heavens must be corruptible. If someone should say that the lack of corruption in the heavens is something acquired from a separate, eternal power, this would mean that something possible with respect to corruption can become eternal. That this is impossible, however, was stated at the end of the first book of *De Caelo*.

189. *De Caelo* I, 2, 268b, 26-29. In *LDC* II, Comm. 71, 146 CD, Averroes states the distinction between the two types of potentialities in the celestial bodies in a slightly different way from what follows in our text. In *LDC*, he distinguishes between potentialities with respect to quality, and potentialities with respect to time. All the celestial

bodies have motions finite in velocity, and fixed in their velocity. That is, each body has a velocity appropriate to it, but all the heavenly bodies are infinite in time.

190. Cf. *LM* XII, Comm. 41, 324 C-E (Arabic: p. 1629, 1-15); *LP* VIII, Comm. 79, 427 BC. For a different resolution of the problem of finite body and infinite force, see *ibid.*, 427 D-G. Aristotle, too, does not speak of eternal motion, but of existence and duration.

191. *Physics* IV, 14, 222b, 30-223a, 15; cf. *ibid.*, VI, 3, 234a, 24-31.

192. See *LDC* II, Comm. 71, 146 C-F; *DSO* III, pp. 109-110 (Latin: 9 M-10 A; Hebrew: p. 41, 87-96); *ibid.*, III, pp. 107-08 (Latin: 9 G-H; Hebrew: pp. 39, 61-40, 69). See also Wolfson, *Crescas*, pp. 612-13.

QUESTION VIII

1. Steinschneider (*HU*, p. 181) tentatively identified this *Question* with the treatise mentioned in the Escurial list under the title, "The Discourse Concerning the Prime Mover" (see Renan, *Averroès*, p. 464). That work, however, is probably to be identified with a lost treatise on the Prime Mover (see Wolfson, "Lost Treatise," p. 683).

2. I.e., it is eternal, not generated and corrupted.

3. Locomotion is itself a form of change, so here we have a general denial of change (*metabolê*), followed by the specific denial of motion in place (*phora*).
IP VIII, v, 4, fol. 136a, 1-9:

> Inasmuch as it had already been demonstrated that there must be an eternal, continuous motion which is prior to the rest of the motions; and because it was demonstrated here that this motion cannot possibly exist in any of the genera of motion except circular translation: it is clear that circular translation must necessarily be a continuous, perpetual, eternal motion. And [because] the body which is moved circularly in perpetuity is the celestial body, it is clear that the mover which causes this motion is the eternal mover whose existence has been demonstrated. It has already been demonstrated from this book, therefore, which motion is the motion that must be perpetual, and [it has been demonstrated] that the mover which causes this motion is eternal, and not subject to motion either essentially or accidentally.

All the forms of change are specifically ruled out at *LP* VIII, Comm. 46, 386 HI:

> Et intendit per hoc, quod dixit: absolutum ab omni transmuta-

> tione, quod non accidit ei aliquis modus transmutationis, nec generatio, nec corruptio, nec alteratio, nec augmentatio, nec diminutio, nec motus localis.

The *locus* in Aristotle is *Physics* VIII, 9, 266a, 6-9; cf. Simplicius, *ad loc.*, p. 1320.

4. The emendation, required by the sense, is supported by a parallel text, *EM* IV, 385 HI (Arabic: p. 128, 3-8; German: p. 106):

> Et, cum fuerit ostensum quòd datur motus aeternus: sed cùm impossibile sit, quòd inveniatur aliquis motus aeternus praeter localem circularem, ut in scientia Naturali fuit declaratum: sequitur igitur manifeste ex hoc qui datur motus localis aeternus. *sed nihil percipitur sensu ita se habere praeter motum corporis coelestis*: motus igitur huius corporis coelestis est aeternus, et eius motor est motor aeternus cuius esse fuit probatum.

5. *Physics* VIII, 10

6. From what follows, it appears that **infinite** is used here in the sense of "infinite in duration." Cf. *Physics* VIII, 10, 266a, 12-23, and below, paragraph 4.
The actual propositions in Aristotle read as follows: (1) "... that nothing finite can cause motion during infinite time," and (2) "... that in no case is it possible for an infinite force to reside in a finite magnitude." On Averroes' understanding of the first of these propositions, see above, *Question* VII, n. 154. In what follows in this text, Averroes will again offer more an interpretation of *Physics* VIII, 10, than a paraphrase of that text.

7. It seems clear to me that something has dropped out of the text at this point. Narboni, however, accepts the reading of our MSS and comments:

> And the second principle, which proceeds from the first, is that it follows that that infinite action belongs to a force which does not inhere in a body at all. That is, because this action is from an infinite force, [and] because the infinite force does not inhere in a body, as has been demonstrated.

8. Cf. *LM* XII, Comm. 41, 325 CD (Arabic: p. 1634, 7-11).

9. For another analysis of Aristotle's proof in *Physics* VIII, 10, see *LP* VIII, Comm. 86, 434 I-M. See also above, *Question* VII, n. 157.

10. I.e., the diurnal sphere.

11. See, e.g., *LDC* I, Comm. 22, 17 LM.

12. *EP* VIII, p. 146, 12-15 (Arabic):

> We say that inasmuch as it has been demonstrated that it [the Prime Mover] has no inherence at all in a subject--which is contrary to the way the case stands with the souls of animals--and that the subject's existence is through the Prime Mover, not the Prime Mover's existence through the subject, it has also been demonstrated that the Prime Mover is not moved accidentally in this way.

Cf. *Question* VII, n. 157, above.

13. Note that while the Greek text of *Physics* VIII, 10, 266a, 18 reads simply *en pleioni gar to meizon* ("for the greater the amount moved, the longer the time occupied" [Ross]), the Arabic version (p. 925, 3) reads: "since it is the greater, it is necessary, therefore, that it should cause motion for a longer time" (*i<u>dh</u> kâna al-'a'zam fa-wâjaba 'i<u>dh</u>an 'an yuḥarrika fî zamân 'aṭwal*). Cf. *Physics* VII, 5 (*LP* VIII, Comm. 78, 424 C: ... *et quod maius est movet tempore longiori. haec enim propositio verificata est in pluribus locis, ut in fine Septimo.*).

14. Narboni, *ad loc.*:

> And his expression **and the second stating that whatever force we originally assume, we can then assume, etc.**, means that whatever finite force we originally assume, we can then assume, etc. It is obvious that this is a defect which has occurred in the translation.

Steinschneider (*HU*, p. 178) cites this comment as part of the evidence for his assertion that Narboni is certainly not the translator of the *Questions*.

15. Cf. Simplicius on *Physics* VIII, 10, 266a, 10 (p. 1321, 18-37).

16. I.e., having assumed initially that there is a finite force which causes motion for an infinite time.

17. Cf. *LP* VIII, Comm. 78, and *IP* VIII, vi, 1, where Averroes reproduces Aristotle's argument, rather than offering an interpretation of that argument as he does here. Cf. also *Question* VII, paragraphs 42-43.

18. *Prior Analytics* I, 15; cf. *De Caelo* I, 12, 281b, 2-15, and *Metaphysics* VIII, 4, 1047b, 9-14.
IDC I, x, 2, 3, fol. 171a, 19-171b, 11 (Latin: 289 BC):

> ... and he said there are four things distinct from one another in name and definition: impossible, false, possible, and true. The possible is [e.g.] when we say that it is possible for the sides of a triangle to be equal. The impossible is to say that its hypoteneuse is equal to a side. The false [e.g.] is when we

say, pointing to Reuben, that he is standing when he is [actually] sitting. The distinction between the false and the impossible is that it is possible for the false to become true, but this is not possible for what is impossible.... The second principle is what was demonstrated in *Prior Analytics*: what is impossible cannot follow from the possible falsehood, but only from the impossible. I.e., when the conclusion of the syllogism is impossible, there must be some impossible falsehood in the premises, not [merely] a possible falsehood.

EDC I, pp. 33, 15-34, 8:

Aristotle demonstrated by the method of contradiction that an absurdity follows from this, and that after he posited two principles for that purpose. One, that "the false" is said in two ways: the first, when what one assumes is impossible with respect to existence, which is the "absurd falsehood"; while the second is the possible falsehood, i.e. when one assumes that what is not existing does exist, but it is [nonetheless] possible for it to exist: e.g., if you say that Zaid is in the market when he is not actually there. For this, even if it is false, is possible.... This being the case, the nature of the impossibly false is different from the nature of the possibly false, and what follows from the one is necessarily different from what follows from the other.... The second principle is what is demonstrated in *Prior Analytics*: that from the possibly false, no impossibly false [conclusion] follows. Rather, when the existence of the impossibly false follows from something, then [that something is itself] impossibly false.

Cf. *LP* VIII, Comm. 36, 375 KL, and *IPrA* I, cap. 15, 44 I-45 A.

19. *KMH*, *Heqqêš*, p. 22b, 15-16 (Latin: *in Libros Pr. Analiticorum*, 50 H):

The falsehood does not follow from the correct proposition, but follows only from the proposition in the syllogism concerning which there is ground for doubt.

20. See *De Interpretatione* VII, 17b, 16-25; *Int. De Interpretatione* II [23-24], pp. 71, 6-72, 13 Kassem.

21. See Wolfson, *Crescas*, p. 457.

22. This is the explanation of Narboni, *ad loc.*

23. *EP* VIII, pp. 147, 18-148, 2 (Arabic):

... rather, it follows necessarily that if a certain thing in motion is moved by two different forces over one and the same distance, the relation between the times of the motions is the relation between the motive forces.

It should be noted that Averroes, following *Physics* VIII, 10, 266a, 24 ff., is no longer discussing a force which produces an action infinite in duration, but a force which produces an action of infinite intensity, as in *LP* VIII, Comm. 79, 427 F:

> Unde, si dixerimus potentias esse in corporibus, tunc actio earum erit proportionalis, scilicet quod proportio potentiae motivae ad potentiam motivam est sicut proportio velocitatis motus ad velocitatem.

24. Narboni, *ad loc.*:

> This means, if we assume one of these forces to be finite, i.e. in its magnitude, and the other to be infinite in its substance, i.e. in quantity; for it is described as "finite" or "infinite" in so far as it is a force in a body.

MS. Munich 31 adds the words "in its magnitude" to the text of the *Question*, but the word is missing from Narboni's lemma.

25. Cf. *LP* VII, Comm. 35, 335 DE:

> Motores autem...in materia, si aliquis eorum haberet potentiam infinitam, contingeret ut motus esset in non tempore. et haec declarabuntur in ultimo Octavi libri. et Arist. non loquitur hic, nisi de motoribus materialibus. quoniam ista perscrutatio est utilis in dispositionibus motorum non materialem....

26. The thrust of the argument is clear, even though the text is not. Cf., however, the quotation from *IP* in n. 28, below.

27. Reading with *Physics* VIII, 10, 266b, 2 (*dâ'iman* in the Arabic version [p. 926, 9]).

28. *IP* VIII, vi, 2, fol. 137b, 2-9:

> After this has been settled, we assume that an infinite force does exist throughout a finite body or within a finite body, and that this motive force is signified by the letter A. We also assume a certain thing in motion which is moved by that force over a certain distance, and that thing in motion is signified by the letter B, while the distance is signified by C. Then I say that one of two things must necessarily follow from this, and both of them are impossible: either the infinite motive force A moves B over distance C in no time, that is, in an instant; or the finite and the infinite move one and the same thing over one and the same distance in one and the same time.

Cf. *De Caelo* I, 6, 273b, 29-274a, 18; *LDC* I, Comm. 64, 45 A-E:

> Si ergo posuimus alterum agentium finitum, et alterum infinitum, continget ut infinitum agat in nullo tempore quapropter nullam proportionem habebunt in actione cum nullam habeant in quanti-

tate. Et, cum posuimus ipsum moveri (*sic*) in tempore. scilicet infinitum movere finitum, possibile est reperiri aliud corpus finitum, quod in illo eodem tempore aget actionem infiniti secundum modum praedictum: et sic igitur inveniemus duo agentia, quorum alterum est finitum et alterum infinitum, quae agent in eadem actione, et in eodem tempore.

29. *EP* VIII, p. 148, 10-11 (Arabic [reading with the Cairo and Madrid MSS]):

It is evident that this absurdity follows only from our assumption of an infinite force in a finite body, and that from which an absurdity follows is itself absurd; for what we assumed from the discourse is a possible falsehood, and no absurdity follows from the possible falsehood, but only from the impossible falsehood, as has been demonstrated in *Prior Analytics*. The discourse concerning the genus of this syllogism, to the effect that there is no defect in it, has already been given.

See also *IPrA* I, cap. 15, 45 CD.

Cf. Simplicius on *Physics* VIII, 10, 266a, 24 (pp. 1323, 31-1325, 7).

30. See above, *Question* VII, n. 158.

31. The direct syllogisms are perfect syllogisms which require no reduction (see C. von Prantl, *Geschichte der Logik im Abendlande* [repr. Leipzig, 1927] I, p. 700 [cf. II, p. 200, notes]). The terms "direct" and "indirect" are also applied to the method of proving syllogisms (I. Bochenski, *A History of Formal Logic* [New York, 1970], pp. 76-78). However, in *IPrA* I, cap. 11, 132 L Averroes states that everything that can be demonstrated by the categorical syllogism which is called "direct" can also be demonstrated, with the same propositions, *per syllogismum contrarii*. And note that earlier (*ibid.*, 132 H), he has stated, speaking of *syllogismi ad impossibile*: *planum etiam qui haec syllogismi species componitur ex conditionali et praedicativo*. Cf. *KMH*, *Heqqêš*, p. 22b (quoted above, *Question* VII, n. 158).

32. *IDC* II, iii, 7, fol. 211b, 12-22 (Latin: 312 HI):

For this reason, it does not harm our speculation if we consider it to be generated, even though we see that it is impossible and false that it should be generated. But we do not posit the impossible *qua* impossible; we only posit it *qua* possible. Thus, no absurdity results to us by using it with regard to this conception and [its] verification. The philosophers frequently used it in similar speculations. We have already spoken about its use in many places.

Ibid., III, iii, 2, 1, fol. 217a, 7-22 (Latin: 315 E-G):

If so, the possible [proposition] in this syllogism is only assumed *qua* possible, not *qua* impossible: i.e., "the heavy"

> here is only assumed to be susceptible of division because it is heavy, not because it is fire or earth. When the possible is assumed *qua* possible, no impossibility results from assuming it. When, in this preceding syllogism, a possible proposition was assumed *qua* possible, namely, that the heavy *qua* heavy is susceptible of division, and a proposition of undoubted truth was joined to it, namely that the point is not susceptible of division, it followed that the point is not heavy, nor is it light. If it were either heavy or light, an impossibility would follow from this, i.e. that the point would be susceptible of division, and the impossibility does not occur because the possible was posited *qua* possible. Thus, the only remaining alternative is that, in this discourse, it results from our assuming that the point is either heavy or light. Proofs similar to these are what Aristotle frequently made use of in this science. Thus, it is also evident how Themistius erred: for he thought that at this point Aristotle employed a possible premise and deduced a necessary conclusion.

EP VI, pp. 28a, 27-28b, 2 (Arabic: p. 93, 1-6):

> Since this is so, what the preceding discourse assumed to be possible, and then an impossibility followed, *is* possible essentially but impossible accidentally. It is only assumed in so far as it is possible, not in so far as it is impossible, and in this way we assume it together with the dubious proposition in a *demonstratio per impossibile*, so that the impossible follows from it. But when what is possible is assumed to exist, an impossibility does not follow from it, and Aristotle uses many demonstrations of this type...

Cf. *DSO* VI, pp. 127-128 (Latin: 12 AB; Hebrew: p. 54, 44-49); *LP* IV, Comm. 72, 163 I (where Plato is also credited with the use of this type of demonstration).

33. See *LP* VIII, Comm. 78, 424 K-M (the reference there to an explanation *in nostris quaestionibus*, most probably means *QLog*; see notes 47 and 48, below.

34. MS. New York, JTSA 2311 breaks off following the word **if**, and does not resume again until **in the same way** (paragraph 11). One entire leaf is missing.

35. *EP* VIII, p. 147, 12-15 (Arabic):

> This is clear in the following way: because every corporeal force is of necessity subject to division when the body is divided. The meaning of this is that the larger body always has a larger force, and the smaller body always has a smaller force, if both bodies are of one species.

IP VIII, vi, 2, fol. 137a, 19-23:

> ... for with respect to the bodies or the forces which are in bodies, it is evident that they are subject to division through the division of the bodies; and everything which has a greater body has a greater motion, or the motive force in it is greater, provided these things are of the same species. This is clear with respect to both motive bodies and the motive forces which are in bodies.

36. Narboni, *ad loc.*, explains that this refers to the assumption made in the first demonstration (paragraph 4, above).

37. See *DSO* VI, pp. 127-130 (Hebrew: pp. 54, 44-55, 70; Latin: 12 B-E).

38. According to Narboni, this refers to the second demonstration (above, paragraph 5).

39. Note the parallel argument at *LP* VIII, Comm. 79, 426 H, where the subject is not the motion of the sphere but its existence.

40. I.e., that Aristotle had not demonstrated in a valid way that the force which moves the heavens is not inherent in a body. It is worth noting that at *LP* VIII, Comm. 78, 424 K-M, it is not only Avicenna and his followers who are criticized for not understanding Aristotle's demonstrations, but Galen and Alfarabi, as well.

41. Narboni, *ad loc.*:

> **That there is no being which is not a body, subsisting** (*'ômêd*) **in itself, separate from the [celestial] bodies, which is itself a principle of those bodies**: this means that every body acts on something and is acted upon by something else. And they considered it possible for the eternal body to be compounded of matter and form.... The sage R. Abraham b. Ezra said that many thought that the sphere of the Zodiac [fixed stars] was God, because it is not a body. He meant by this, because it is not in place like the things enclosed by place (i.e., [enclosed by]the final sphere), because it encloses and is not enclosed in itself. How much better was Averroes when he said **this is the worst of doctrines, etc....** He means that it is [an opinion] in opposition to all the fruits of philosophy, i.e. the belief in the Separate Intelligence. And it was to this purpose that Aristotle roused and exerted himself, and discovered and ate from the fruit of the tree of life. Averroes devoted a special treatise to this topic...

On the opinion of Avicenna and his followers, see *TAT* X, p. 254 (Arabic: p. 421, 4-10; Latin: 104 M-105 A). On the "Oriental" or "illuminative" philosophy, see C. Nallino, "Filosofia 'orientale' od 'illuminativa' d'Avicenna?" *Rivista degli Studi Orientali* X (1923-25), pp.

433-67; S. Pines, "La 'Philosophie orientale' d'Avicenne et sa polémique contre les Bagdadiens," *AHDLMA* XIX (1952), pp. 5-37; H. Brown, "Avicenna and the Christian Philosophers in Bagdad," in *Islamic Philosophy and the Classical Tradition* (Oxford, 1972; ed. S. Stern), pp. 35-47; and Gutas, *Avicenna*, pp. 115-130. For a different approach, see H. Corbin, *Avicenna and the Visionary Recital* (New York, 1960), pp. 35 ff.

42. On the basis of the passage from *TAT* cited in the preceding note, one might think that **the philosopher** (*he-ḥâkâm*) refers to Avicenna. Averroes, however, is quite consistent in reserving this epithet for Aristotle. See, e.g., *Question* I, paragraph 1; Index A, b to *LM* (Bouyges), *s.v. al-ḥakîm*.

43. Cf. the reference to the oral teaching of the ancient philosophers above, *Question* III, paragraph 12. Perhaps this interpretation of the work of the ancient philosophers is connected with the rise of Bâtiniyya movements (see Peters, *Aristotle and the Arabs: The Aristotelian Tradition in Islam* [New York, 1968] , p. 177). In any case, it is interesting to find Averroes himself using terminology appropriate to the interpretation of the Quran with respect to Aristotle: e.g., *huwa al-ẓâhir min al-madhab 'Arisṭû* (*EM*, p. 152, 12-13 [Arabic]; cf., e.g., *ICat* II, iv, p. 88, 3 [Bouygès]; p. 131, 7-8 [Kassem]).

44. The text and the sense are both difficult. The reading of MS. Munich 31 is unclear, but may be *h.m.d.h.* This may suggest that the original wording of our text may have been "because of error and ignorance."

I cannot explain why Avicenna's followers should have thought (or should have been thought to think) it necessary for Aristotle to conceal a belief that the mover of the celestial sphere inhered in that sphere, a belief which Plato had already stated. One might think they meant to imply that Aristotle, too, held that God was not the Prime Mover. Our text, however, states that Avicenna denied the existence of any incorporeal being itself the principle (=cause) of corporeal *existence*, which amounts to something more than the denial of an incorporeal motive cause. Such a position is reminiscent of the Dahriyya. (Perhaps the term *ha-siklût* [=ignorance] in our text is equivalent to the Arabic *al-jâhiliyya*, which also means ignorance, and has specific reference to the barbarous period before Islam, the period to which the existence of the Dahriyya is ascribed [see H. Wolfson, *Repercussions of the Kalam in Jewish Philosophy* (Cambridge, Mass. and London, 1979), pp. 153-59].) If Avicenna did hold such a position (not found in any of his extant works), it must certainly have been expressed in an esoteric text (but see the passage in *TAT* cited in n. 41, above).

45. **Newfangled** is an attempt to preserve the reading *berû'îm* which is found in both MSS. It is, however, not unlikely that this reading is a corruption of *bedûyîm* = false.

46. Cf. *IPrA* I, cap. xv, 44 I-M.

47. Cf. *Metaphysics* VIII, 4, 1047b, 14-30; *QLog* VII (*libri priorum*), 93 EF.

48. I.e., with respect to both the sublunary and the celestial bodies. Cf. *QLog* VII (*libri priorum*), 95 F.

49. See paragraph 4, above.

50. *De Caelo* I, 9, 278b, 11-279a, 11; on assumptions made only in so far as they are possible, see *LDC* II, Comm. 104, 167 A; and n. 32, above. Cf. *LP* VII, Comm. 2, 307 LM; and n. 56, below.

51. See paragraph 5, above.

52. I.e., time and motion are continuous quantities; see *Physics* VI, 2.

EP VI, p. 26b, 12-14 (Arabic: p. 88, 1-3):

> When we have assumed these two propositions, it follows from them that the continuous *qua* continuous is divisible into what is always divisible.

IP VII, i, fol. 102a, 7-8: "...inasmuch as it has already been demonstrated that everything in motion is divisible into what is always divisible."

See also *LP* VI, Comm. 3, 248 E; *ibid*., Comm. 15, 255 E: ...*et est quod, ex hoc, quod velocius et tardius pertranseunt in infinitum, sequitur quod magnitudo dividetur in divisibile.*

53. *IP* VI, i, 3, fol. 86b, 15-23:

> A proof of this [that there is no indivisible time or magnitude] is that we have already assumed that the faster is moved over one and the same distance in a shorter time than that in which the slower is moved, and that the faster is also moved in one and the same time over a greater distance than that over which the slower is moved. If this is so, and we assume that a certain thing in motion is moved over a certain distance in a certain time, it is possible for us to assume a thing in motion which is faster and one that is slower, in so far as what is moved [is moved] in time. The faster necessarily divides the time, and the slower divides the magnitude, for the slower is moved in this time over a part of this magnitude, and the faster is moved over this magnitude in less than this time.

LP VI, Comm. 16, 256 H-M:

> Cum posuit quod velocius pertransit magnitudinem, quam pertransit tardius in minori tempore, et quod tardius pertransit in illo tempore, in quo pertransit velocius magnitudinem minorem magnitudinem, quam pertransit velocius, et quod velocitas, et tarditas procedunt in infinitum, conclusit ex hoc quod divisi-

bilitas in magnitudine, et tempore procedit per divisionem temporis à velociori, et magnitudinis à tardiori in infinitum. et, cum divisio est in eis semper: et hoc est, definitio continui: necesse est ut magnitudo, et tempus sint continua....

54. I.e., in an instant.

55. *EP* VI, p. 27b, 22-28 (Arabic: p. 91, 4-9):

> We say that it must necessarily also be true with respect to time that it is not composed of indivisibles, for if we assume an indivisible time in which a certain thing in motion is moved over a certain distance, we cannot assume a certain thing in motion over the same distance which is faster than the first: for if we did make that assumption, it would follow that the faster would traverse that same distance in a shorter time than the time we assumed to be indivisible.

Cf. *LP* VI, Comm. 11, 252 H-253 B.

56. *Physics* VII, 1, 241b, 24-242a, 15; cf. above, *Questions* I and VII.
EP VII, pp. 35b, 24-36a, 11 (Arabic: pp. 115, 17-116, 12):

> The truth of the propositions which Aristotle employed with respect to the proof of this problem had already been demonstrated from this discourse, and he demonstrated that everything in motion, in so far as it is in motion, has a mover. But a certain difficulty has arisen with respect to this discourse, namely, that that thing which is in motion can come to rest when a part of it [comes to rest] only if it is assumed that rest is possible for that thing. If it is demonstrated that there is something for which rest is impossible, as Aristotle thought concerning the celestial bodies, then I do not know how what is not possible can be assumed to be possible in this proof! If this is true, and rest is impossible for some of the things in motion, this demonstration is particular and it cannot be truly concluded from it that everything in motion has a mover, so that it ought to be considered an illusory proof rather than a demonstration. But we say that if there is some body in motion for which rest is impossible, then this is impossible for it not because it is in motion, but because it is something in motion of a particular description: e.g., because its mover is eternal, or because it has no contrary; but in so far as it is in motion, it is possible for it to come to rest. Consequently, the possible was assumed to be possible here only in so far as it is possible, not in so far as it is impossible, [and hence] the aforementioned demonstration follows from it. Many of the demonstrations employed in mathematics are of a similar nature, and no error occurs because of their use.

LP VII, Comm. 2, 307 K-308 C:

> Declaratum est igitur ex hoc sermone quod ista demonstratio est vera, et quod est de genere signorum certorum: licet not sit de genere demonstrationum simplicium. Et potest quis dicere quod propositiones, quae usitatae sunt in ea, non sunt utiles omne moto ex se. cum in corpore coelesti impossibile est dicere quod, cum quieverit pars, quiescit totum, cum imaginari hoc sit impossibile. ista autem demonstratio fundata est super hoc, quod partem quiescere in omni moto est possibile. quod est impossibile in corpore coelesti....Ad hoc autem dicendum est, quod partem quiescere corporum, et de quibus demonstratium est, quoniam impossibile est ut corrumpantur, aut quiescant, quodammodo est possibile, quodammodo impossibile. est enim possibile, secundum quod est corpus, impossibile est autem, secundum quod est aliquid corpus, v.g. neque leve, neque grave. et, quia consyderatio haec est de corpore, secundum quod est corpus, cum motum, secundum quod est motum, est corpus. ideo illud quod positum est hic, est possibile. Et sciendum est qui multa necessaria sequuntur per se multa possibilia: et quòd propter hoc est possibile in signis, ut propositiones eorum sint compositae ex possibilibus, et necessariis: sed tamen necesse est ut maiores sint necessariae: quoniam hoco modo contingit quòd conclusio sit necessaria....

For another objection to the argument of *Physics* VII, 1, which is answered in the same way, see Simplicius, *in Physicorum*, pp. 1039, 13-1041, 25.)

Abravanel, *Šâmayim Ḫadâšîm* III, p. 29a, 11-21:

> You cannot say that what I have assumed, to the effect that the sphere is not in motion, is an impossible assumption, and for that reason this impossibility follows from it. For it is possible to imagine the body *qua* body to be at rest, and it is possible to imagine it to be in motion, in spite of the fact that in so far as it is a certain kind of body it is in motion perpetually and does not rest. But an impossible falsehood does not follow from a possible falsehood. Aristotle had already employed a similar type of demonstration with respect to this discipline in the sixth book of the *Physics*, when he assumed the possibility of a motion faster than the diurnal motion and the possibility of a body greater than the encompassing sphere. In his commentary [to *Physics* VI], Averroes wrote that Aristotle did not err, nor did he intend to cause others to err with respect to this matter, and that what he deduced from this assumption is a demonstrative conclusion. For the heavens, in so far as they are bodies, could be moved with a faster motion; and the sphere, in so far as it is a body, could be imagined to be greater than it is. All this is possible with respect to the heavens in so far as they are a body, in spite of the fact that in so far as they are a certain kind of body this is not possible with respect to them. And [Averroes also wrote] that

> Aristotle assumed these things in so far as they are possible, and Averroes' argument on this is correct. I say, therefore, that the sphere, in so far as it is a body, may be imagined to be at rest and not in motion...

Cf. *LP* VI, Comm. 15, 255 I-L; and *EP* VII, p. 116, 13-15 (Arabic):

> It is amazing that Abu Bakr [Ibn Bajja] solved this difficulty here in a manner similar to our solution, but he did not do this in the sixth book when he said that there is not something faster, or something slower, than anything in motion, so that he deviated from Aristotle's proof...

See also *DSO* VI, p. 131 (Latin: 12 HI; Hebrew: p. 56, 84-93).
It is instructive to compare this *Question* with Simplicius' treatment of *Physics* VIII, 10.

QUESTION IX

1. In his "Filosofia 'orientale,'" p. 437, n.1, Nallino identifies a treatise mentioned in Ibn Abi Uṣaybi'a's list under the title, "A treatise refuting Ibn Sina's division of existing things into merely possible; possible *per se*, necessary through something else; and necessary *per se*," with a lost work of Averroes'. (See Renan, *Averroès*, p. 456 [reading with the MSS, rather than with Renan; cf. *ibid*., p. 458].) That treatise, however, is to be identified with this *Question*, which in the MSS of Ṭodrosi's translation is introduced by the words: "A Treatise of Ibn Rushd's criticizing Ibn Sina's division of existing things into possible *per se*; possible *per se*, necessary through something else; and necessary *per se*." (See also Steinschneider, *HU*, pp. 181-182.) It should be noted that the Escurial list (Renan, *op. cit*., p. 464), which lacks the title mentioned above, does contain the following title: "A treatise concerning the necessity of the eternal [*thâbita*] body in its rotation."
Averroes' other writings contain many parallels to the contents of this *Question*. Ṭodros Ṭodrosi, in the introduction (found only in MSS. P. 1023 and BM Add 27559) to his translation, refers the reader to the end of *LP* VIII (see Comm. 79, 426 L and 427 D). More extensive parallels are to be found in *DSO* VII, *TAT*, and, above all, in *IDC* I, x, 2, 8.
To my knowledge, the edition of this work promised by Worms (*Lehre*, p. 37) never appeared.

2. See Avicenna, *al-Najât* (Rome, 1593), Metaphysics II, pp. 62 ff.; M. Horten, *Die Metaphysik Avicennas* (Halle, 1907-09), I Abh., ch. 6, pp. 62-77, and VIII Abh., ch. 4, pp. 498-506 (and for a translation of the same passages, based on a later edition of the Arabic text, *Avicenne: La Métaphysique du Shifâ'*, trans. G. Anawati [Paris, 1978/1985], vol. I, pp. 113-116; vol II, pp. 85-89); T. Haarbrücker, *Asch-Schahrastaniis Religionspartheien und Philosophenschulen* (Halle, 1850-51), System des Ibn Sina II, pp. 236-37, and 250 ff.; P. Morewedge, *The Metaphysica of*

Avicenna (New York, 1973), pp. 47-50. Note that "possibility" is cited as one of the subjects treated in the *Oriental Philosophy* (see Pines, "'Philosophie orientale'," p. 26; cf. above, *Question* VIII).

Elsewhere Averroes claimed that Avicenna took this classification of being, and the proof for the existence of the First Principle based on this classification, from the Mutakallimûn (see *TAT* IV, pp. 163-64, and van den Bergh's n. 72. 1 [Arabic: pp. 276, 1-277, 11; Latin: 71 K-72 B]; cf. above, *Question* V, where certain related arguments are cited not in the name of Avicenna, but in the name of the Mutakallimûn. Cf. also Wolfson, "Lost Treatise," pp. 697-99.) The validity of the metaphysical nature of Avicenna's proof for the existence of the First Principle is not considered in this *Question*, nor does Averroes raise the problem of whether the **necessary of existence** differs from the Prime Mover.

3. But cf. *LP* VIII, Comm. 79, 426 LM:

> Audiens autem Avi. verba haec Ari. qui iam audierat verba Alex. opinatur est duplex esse necessarium, necessarium scilicet ex altero contingens, et possibile ex seipso, et necessario ex se: necessario quidem ex alio, ut Coelum, necessario ex se, ut motores Coeli.

Cf. also below, paragraph 5.

On Alexander's responsibility for Avicenna's error, see also below, n. 38.

4. For the construction of the phrase **necessary of existence**, see Wolfson, *op. cit.*, p. 685, n. 8. Averroes defines Avicenna's phrase "entity of a necessary existence" as an "existence without cause" at *TAT* IV, p. 164 (Arabic: p. 277, 3-4; Latin: 71 M).

H. Davidson has used the three-fold division of existence attributed here to Avicenna as one piece of evidence showing that Averroes' knowledge of Avicenna's writings was deficient (*Proofs*, ch. X, pp. 318-20; 334-35; Avicenna speaks of two types of being). While Professor Davidson's conclusion may be perfectly correct, it seems to me that the three-fold division found here is of limited evidentiary value, for three reasons. First, the polemic context of the work; Averroes is not noted for fairness to opponents. Second, the fact that, as Professor Davidson himself notes, all interest here is focused on the impossibility of a category possible in itself, necessary through another. Third, the fact that Averroes is very fond of three-part divisions in general, sometimes turning an *Aristotelian* two-part division into a three-part division (see Hugonnard-Roche, "Méthodes," pp. 247-51). Additionally, the three-fold division found in this *Question* is not the rule in Averroes' writings; see e.g., the passages from *TAT* and *LP* cited above, nn. 3-4. Throughout *TAT*, for example, Averrroes never refers to the first part of the tripartite division (the merely possible--the part of the classification that most excites Professor Davidson's wrath), at all. It might be possible to argue that Averroes, over the course of time, improved his acquaintance with Avicenna's writings, but I would find it difficult to believe that this *Question* is an early work, even granting the difficulties of dating the various treatises.

Given the context of this treatise, I would conjecture that Averroes is simply attempting to set the case of generated and corrupted beings aside. Admittedly, he does this by using the term "possible" in quite different way from the way Avicenna uses this term. But one may question whether this is due to ignorance or to Averroes' insistance on using terms like "possible" and "necessary" in what he perceives to be Aristotle's sense. It is, after all, one of Averroes' complaints against Avicenna that the latter has strayed from the true Peripatetic tradition.

However, Avicenna himself recognized a difference between generated and corrupted beings and the eternal heavens with respect to possibility and necessity (see F. Rahman, "The Eternity of the World and the Heavenly Bodies in Post-Avicennan Philosophy," *Essays on Islamic Philosophy and Science*, ed. G. Hourani [Albany, 1975], pp. 222-23; A. Ivry, "Destiny Revisited: Avicenna's Concept of Determinism," *Islamic Theology and Philosophy*, pp.162-67).

5. Cf. Avicenna, *al-Najât*, Metaphysics I, p. 60: "... and the essentially eternal is that which has no principle [=cause] for its essence through which it exists...".

Aristotle identifies the eternal and the necessary at *De Gen. et Corr.* II, 11, 338a, 1. On this identification, see Wolfson, *Crescas*, pp. 109 ff., and 680 ff.

6. See *TAT* VIII, p. 238 (Arabic: p. 395, 1-15; Latin: 99 D-F); *ibid.*, X, p. 252 (Arabic: pp. 417, 13-418, 15; Latin: 104 C-F). Averroes is, however, prepared to accept this classification of being *in toto* if it is stipulated that it does not actually exist in things, but is a result of an operation of the intellect. In the Hebrew MS of *DSO* VII (P. 989, fol. 29a-b, where it immediately follows the text of this *Question* and bears the title: "A treatise of Averroes' exculpating Ibn Sina about that for which he criticized him in the preceding treatise"), Averroes compares this operation of the intellect to the task of the geometer, and then (fol. 29b, 1-4; Latin: 14 FG) goes on to say:

> The discourse of Avicenna is also true, namely, that the compelled things are of two classes: those which are compelled by themselves and necessary in virtue of their own essence, and this class is called "necessary of existence in virtue of itself"; and those which are compelled through something other than themselves [but] possible of themselves, that is considered in themselves--not that there is [actually] any possibility at all in them. These are the things concerning which it is said that their nature changes from the possible to the compelled--but in reason, not in existence.

(The fact that *DSO* VII follows our treatise, of course, does not mean that *DSO* VII must have been composed later than our treatise.)

Cf. Shem Tob's commentary to *Moreh Nebukim* II, Introduction, Proposition XIX (p. 9a-b):

> Being, according to mental division, is to be divided into nec-

> essary and possible. The necessary is that whose existence is from itself and which has no dependence on any other thing. The possible is that which has existence from something other than itself, and is caused.... Avicenna thought that this distinction was not a mental distinction, but rather that the possibility was something from the mover, and, similarly, this is the opinion of Maimonides, as will be demonstrated in what is to come. Thus, he thought that the thing is possible in virtue of itself and necessary in virtue of its cause... Averroes thought that possibility was not something that existed at all, but that if we call what is caused "possible," it is in virtue of its having a cause. Note this, for it is something profound.

Cf. also Vajda, *Isaac Albalag*, pp. 172-74, and see *Metaphysics* V, 5, 1015b, 9-15 (cf. *LM* V, Comm. 6, 109 K-110 C [Arabic: pp. 520, 11-523, 8]).

7. *IDC* I, x, 2, 8, (4), fol. 177a, 10-13 (Latin: 293 BC):

> If there is a nature capable of eternity, it is impossible that it should be corrupted; similarly, if there is a nature capable of generation and corruption, it is impossible that it should continue eternally --unless it is possible that the nature of the necessary can be converted to the nature of the possible.

See also *EDC* I, p. 36, 6-8; *TAT* II, p. 71 (Arabic: pp. 120, 14-121, 1; Latin: 38 KL); *ibid.*, III, p. 146 (Arabic: p. 246, 5-10; Latin: 65 GH); *LDAn* III, Comm. 5, p. 391, 140-43; *Kašf*, pp. 57-59 (Cairo; Müller: p. 39; and Wolfson, "Lost Treatise," pp. 700-02.

Note that in *TAT* III, al-Farabi, as well as Avicenna, is credited with believing in the conversion of the possible to the necessary.

Note also that at *DSO* II, pp. 82-84 (Hebrew: p. 28, 54-65; Latin: 6 IK), Averroes himself expresses the very view disputed in this *Question* (cf. *LM* II, Comm. 4, 30 CD [Arabic: pp. 14, 15-15, 16]).

For a discussion of Averroes' views on possibility and necessity (without reference to this *Question*), see G. Jalbert, "La nécessité et la contingence chez Aristote et Averroès," *Revue de l'Université d'Ottawa* XXX (1960) pp. 30*-35*.

8. I.e., potentiality for corruption, not for motion in place. See below, paragraph 5.

9. The reference is to *De Caelo* I, 12. For a discussion of the distinction between potentiality and possibility, see Wolfson, *Crescas*, pp. 690-93.

LDC II, Comm. 34, 117 K:

> ...possibile enim in rebus aeternis est necessarium, cum nullum possibile sit in eis vere. Et causa huius est quod nulla potentia est in aeterno ad recipiendum eius contrarium, quod habet...

For other parallels to our text, see, e.g., *LDC* I, Comm. 109, 75 AB; *LM* IX, Comm. 17, 243 K (Arabic: p. 1200, 9-12); *DSO* III, pp. 103-104 (Hebrew: p. 38, 38-44; Latin: 9 CD); *ibid.*, IV, p. 116 (Hebrew: p. 49, 25-26; Latin: 10 GH).

10. See *TAT* V, p. 180 (Arabic: p. 304, 13-16; Latin: 78 A); *ibid.*, VIII, p. 241 (Arabic: pp. 399, 13-400, 7; Latin: 100 EF); *EM* II, 373 A (Arabic: p. 75, 1-6; German: pp. 61-62). See also Wolfson, *Spinoza* I, pp. 122-25.

11. On the fact that "eternal" does not mean "causeless," see above, *Question* V. See also below, the end of paragraph 5, on the corporeal substance of the heavens.

12. Cf. above, *Question* VII, paragraph 48; and below, paragraph 7.

13. Cf. *Question* VII, paragraph 47.

14. See *De Caelo* I, 3, 270a, 12 ff.; cf. *Physics* VIII, 7, 260b, 26-30. See also *LP* I, Comm. 63, 38 F; *LM* XII, Comm. 41, 324 CD (Arabic: p. 1629, 1-10); *ibid.*, Comm. 29, 313 F (Arabic: p. 1557, 9-12); *EM* III, 377 LM (Arabic: p. 95, 1-9; German: p. 79); *TAT* I, p. 34 (Arabic: p. 59, 1-2; Latin: 26 C [with variant reading]); *DSO* II, pp. 79-80 (Hebrew: pp. 26, 27-27, 35; Latin: 6 E); *ibid.*, IV, p. 116 (Hebrew: p. 49, 25-29; Latin: 10 H); *ibid.*, VI, p. 133 (Hebrew: pp. 56, 109-57, 112; Latin: 12 M). Note, too, the interesting passage from *LP* VIII, Comm. 15, 352 B:

> Cum autem perscrutatus est secundum cursum naturalem de istis rebus, vidit hoc esse ens medium inter aeternum, et generabile, et corruptibile: et quod quodam modo est aeternum, et alio modo generabile, et novum: et est corpus, quod movetur circulariter, quod est generabile, et corruptibile quantum ad ubi, et manens quantum ad substantiam.

15. See, e.g., *LM* XII, Comm. 10, 296 M-297 A (Arabic: p. 1447, 4-16); *TAT* IV, pp. 160-61 (Arabic: pp. 270, 11-272, 16; Latin: 70 G-K). See also Wolfson, *Crescas*, pp. 594-98.

16. See *LM* VIII, Comm. 12, 220 FG (Arabic: p. 1077, 4-18); *EM* II, 372 BC (Arabic: p. 71, 9-15; German: p. 58); *DSO* II, p. 82 (Hebrew: p. 27, 47-51; Latin: 6 GH); *ibid.*, III, pp. 110-11 (Hebrew: pp. 41, 100-42, 108; Latin: 10 BC). Cf. *Metaphysics* XII, 2, 1069b, 24-26. The insistence on the term "subject" for the matter of the heavens is also found in Simplicius; see, e.g., *in De Caelo*, p. 134, 5-10.

17. *LP* I, Comm. 79. 45 D:

> Et hoc declarabitur in de substantia orbis quod corpora coelestia non habent materiam omnino, quoniam tunc essent generabilia, et corruptibilia propter mixtionem privationis cum natura [a.l. materia] eorum. nihil aliud enim est causa generationis, et cor-

ruptionis quam prima materia propter non esse, quod mixtum est in substantia eius.

See also *TAT* I, p. 43 (Arabic: pp. 74, 14-75, 2; Latin: 29 F). Cf. *DSO* II, p. 74 (Hebrew: p. 25, 1-5; Latin: 5M; *LM* XII, Comm. 7, 295 HI (Arabic: p. 1438, 7-14).

18. *De Caelo* I, 3, 270a, 12 ff.; but see Simplicius, *op. cit.*, p. 92, 7-32 (cf. p. 108, 2-21), where the proof is ascribed to *Physics* VIII. *IDC* I, vi, fol. 146a, 12-146b, 4 (Latin: 275 D-G):

> After he [Aristotle] had completed this, he began to demonstrate that this noble body is neither generated nor corrupted, and that it is not capable of increase or diminution, or alteration, or being acted upon, and that it is altogether unreceptive of the qualities which are consequent to passive alteration.... He demonstrates this first by means of two propositions: one, that every generated and corrupted thing exists only from its contrary, and is corrupted into its contrary.... The second proposition is that the circular body has no contrary.... The first proposition is both self-evident and evident through induction. [But] in order to prove that the body moved in a circle has no contrary, he again uses two propositions The first is that the contrary elements must necessarily have contrary motions: for these bodies are only described as contrary because of their forms which are the principles of the motions; thus, the motions of contrary things must be contrary. The second proposition is that the circular body has no contrary to its motion. Thus, it follows that the circular body has no contrary.

Ibid., I, vi, fol. 147a, 4-9 (Latin: 275 KL):

> After it had been demonstrated that this [celestial body] has no contrary, [it became clear that] it has a subject which must, therefore, be simple, not compounded of matter and form, and this subject is like the actual matter belonging to abstract forms. But it is more like matter than form, even though it has some resemblance to both: for it is like matter because it is sensible and perceptible, and because it has a potentiality with respect to place, and because it is a body; but it is like form because it is actual and its substance is not potential.

Ibid., I, vi, fol. 149b, 9-12 (Latin: 277 F):

> It is indeed evident that the circular body has no contrariety in it: for the cause of contrariety is "up" and "down"; and the cause of "up" and "down" is the circular body, so that the circular body is the cause of contrariety. Thus, it has no contrariety in it. I have given this proof even though Aristotle does not mention it.

See also *LDC* I, Comm. 18, 13 I; Comm. 20, 15 B-D; Comm. 22, 18 D-F.

19. *De Caelo* I, 12

20. On paragraph 5, see *Metaphysics* IX, 8, 1050b, 16-28. See also *TAT* III, p. 142 (Arabic: p. 239, 1-10; Latin: 63 M-64 A); *ibid.*, p. 145 (Arabic: p. 243, 5-15; Latin: 64 M-65 A). Cf. Simplicius, *op. cit.*, p. 312.

21. See *TAT* X, pp. 254-55 (Arabic: pp. 422, 5-423, 3; Latin: 105 CD). Cf. Wolfson, *Crescas*, p. 588.

22. Cf. *LP* VIII, Comm. 83, 432 CD.

23. See *LP* VIII, Comm. 78, 424 H-K; *DSO* I, pp. 72-73 (Hebrew: p. 24, 189-193; Latin: 5 I-K); *TAT* III, p. 141 (Arabic: p. 236, 6-14; Latin: 63 EF).

24. Cf. *Metaphysics* XII, 7, 1072b, 4-14.

25. *Physics* VIII, 1, 251a, 8-252a, 5

26. *IP* VIII, ii, 5, fol. 115b, 19-20: "Thus, it has already been demonstrated from this discourse that motion, as a genus, does not fail." See also *LP* VIII, Comm. 9, 345 F (cf. *ibid.*, Comm. 36, 375 L). But cf. *EP* VIII, p. 40a, 24-28 (Arabic: p. 133, 5-8):

> This is the suitable explanation of the discourse of Aristotle at the beginning of this book, [and] not what certain men thought: that Aristotle's intention was only to demonstrate that motion as a genus does not fail. For Aristotle's speculation [there] concerned only the totality of the world, while the motion which does not fail is in a part of the world.

Cf. also above, *Question* VII, paragraphs 15-18.

27. See *TAT* VIII, p. 237 (Arabic: p. 394, 3-4; Latin: 99 B). Note also *IDGC* II, iv, 2 (68), p. 96, 61-64 (Latin: pp. 159, 20-160, 4).

28. *LP* VIII, Comm. 46, 387 H-K:

> Si igitur dicatur quod non apparet ex istis generabilibus quod sunt aeterna, nisi secundum praeteritum. et demonstratio non complebitur, nisi ponantur aeternal in praeterito, et in futuro. dicemus ad hoc, quoniam, cum declaratum sit quod impossibile est ut deficiant in praeterito infinito, declarabitur quod est impossibile in eis ut deficiant in futuro. quoniam, si deficerent in futuro, deficerent in praeterito infinito. quoniam illud, in quo est potentia infinita, non est determinatum magis ad unum tempus, quam ad aliud. et, si deficerent in praeterito, non essent nunc...

On Aristotle's use of the "principle of plenitude," see R. Sorabji, *Necessity*, pp. 128-132.

29. The Hebrew phrase used here for the **in the uppermost celestial region** is *be-'ôpeq ha-kôl*. *Ha-kôl* (the All) is the equivalent of the Greek *to pan*, a term which usually means "the Universe," but which can be used with reference to the heavens (see *Physics* IV, 5, 212b, 17-18; *De Caelo* I, 9, 278b, 10-22). The term *'ôpeq* normally means "horizon" in Hebrew. Here, however, combined with *ha-kôl*, it must refer to the outermost, or upper, region of the universe (the *eschaton ti tou pantos* of *De Caelo* I, 7, 276a, 7). The same Hebrew terminology is found in *IDC* I, vii, fol. 159a, 23-159b, 4 (Latin: 282 IK):

> A third proof: namely, that the infinite body is not moved with rectilinear motion, nor is it heavy or light. If it were light, it would be in the upper region of the heavens (*be-'ôpeq ha-kôl*), and if it were heavy, it would be in the middle. But what is infinite has no "upper" or "middle."

30. See *De Caelo* II, 1, 284a, 2-11; *LP* VIII, Comm. 1, 339 C and E. I have followed Ṭodrosi's translation which consistently refers to one encompassing motion. The other MSS use plurals here, and return to the singular at the end of paragraph 6.

31. See *DSO* IV, pp. 116-17 (Hebrew: p. 49, 30-38; Latin: 10 HI; *LP* VIII, Comm. 57, 398 G; *TAT* XIV, p. 293 (Arabic: p. 481, 6-8; Latin: 119 EF).

32. See above, *Question* VII, paragraph 45. Averroes does, however, state explicitly that the motion of the heavens is possible in *LM* XII, Comm. 41, 324 I-325 A (Arabic: pp. 1632, 1-1633, 10); *TAT* VIII, pp. 237-238 (Arabic: pp. 394, 11-395, 13; Latin: 99 C-F); cf. *ibid.*, IX, p. 244 (Arabic: p. 405, 3-10; Latin: 101 F), where this is implied.

33. See *LDC* II, Comm. 6, 98 F-H, referring to *De Caelo* II, 1, 284a, 26-32. Cf. *DSO* IV, pp. 112-15 (Hebrew: p. 48, 6-24; Latin: 10 D-G).

34. Cf. *LM* XII, Comm. 30, 315 A-C (Arabic: pp. 1567, 14-1568, 13).

35. Cf. *TAT* IX, p. 244 (Arabic: p. 405, 3-7; Latin: 101 EF).

36. But cf. *LM* XII, Comm. 41, 324 H-325 C (Arabic: pp. 1631, 8-1634, 10).

37. According to *DSO* VII, 14 E (Hebrew: P. 989, fol. 29a, 42-45), Alexander did this in his work *On the Governance of the Spheres* (see F. Rosenthal, "Arabic Manuscripts V," p. 17). Cf. *IDC* I, 293 GH (Hebrew: I, x, 2, 8, fol. 177b, 12-22).

38. It is strange to find Alexander, the author of a work entitled "That the World Cannot be Incorruptible by the Will of God if it is Corruptible in its Own Nature" (*Quaestionum naturalium et moralium ad Aris-*

totelis philosophiam illustrandum [Monachii, 1842; ed. L. Spengel] I, 18), being accused of holding an opinion identical to that of Avicenna on the inherent possibility of the world. Of course, this work was not available to Averroes, and he could not have known that Alexander there defined the possible as that which can or cannot be, depending on whether or not its "becoming" is hindered. Alexander also holds there that the corruptible must be corrupted: if it acquired incorruptibility from some cause, a thing would be corruptible and incorruptible at one and the same time.

But Averroes makes this charge against Alexander in more than one place: e.g., *DSO* VII; *IDC* I, 293 GH; *LP* VIII, Comm. 79. Cf. *DSO* III, where the Hebrew text (pp. 38, 47-39; 51) implies that Alexander did not actually hold this opinion, but that Avicenna misinterpreted what Alexander had written. This reference is lacking in the Latin edition of 1562 (9E), but according to Steinschneider (*HU*, p. 186, n. 580), it does exist in the Latin edition of 1489. We have a similar reference to Avicenna's misinterpretation of Alexander, although on a different topic, in *LM* XII, Comm. 5, 293 K (Arabic: p. 1426, 12-13).

Simplicius brings a charge of another kind against Alexander--again, in more than one place. In his attempt to combat Alexander's argument against Plato's view that that the world is corruptible in its own nature, eternal through the will of God, Simplicius equates Plato's position with that of Aristotle in *Physics* VIII, and wonders why Alexander did not realize the equivalence of these positions. (See, e.g., *in De Caelo*, pp. 301; 352, 24-353, 10; 358, 27-361; *in Physicorum*, pp. 1358 ff. Averroes, on the other hand, thinks that Alexander's *interpretation* of Aristotle is equivalent to Plato's position [*LP* VIII, Comm. 79, 426 H-L].) Simplicius does not, at least in this connection, cite Alexander's work *On the Governance of the Spheres*; presumably, he relies on Alexander's commentaries to *Physics* and *De Caelo*. (Perhaps the treatise, like the "Refutation of Galen," is an excerpt from a commentary; see above, *Question* I, n. 5.)

For Philoponus' disagreement with Alexander, see Simplicius, *in De Caelo*, 78, 12-79, 14 (tr. Wildberg, pp. 65-67).

Although both Averroes and Simplicius for the most part refer to Alexander's "error" in a "*contra Philoponum*" context, it seems much clearer in Simplicius' arguments that Alexander's interpretation of *Physics* VIII applied to the motion of the sphere, not its substance. (For a Neoplatonic statement which appears to me to include the substance, see *Procli Diadochi in Platonis Timaeum Commentaria* [ed. E. Diehl; Leipzig: 1903], vol. I, pp.260, 14-15; 267, 16-268, 6, *et passim*.) It is Alexander's attempt to solve the problem of how a finite body which must have finite force can move eternally, by postulating an infinite receptivity in that body, that enables Simplicius to deduce that the motion of the sphere depends on something else, and then to proceed to exculpate Plato.

That Averroes takes Alexander to refer only to the motion of the sphere is not so clear to me. Indeed, Averroes himself at times held that the sphere is necessary in its substance, possible in its motion (see the end of paragraph 5, above, and the passages cited in n. 32, above; cf. van den Bergh, n. 244.3 to *TAT*). One ought, again, to note that in *DSO* VII, Averroes is willing to admit the validity of the positions of both Alexander and Avicenna, so long as they are taken simply as mental constructs.

Clearly, Averroes' thought on the subject of possibility, as well as his opinion of Alexander, was something that varied considerably throughout his career. Perhaps his insistence in this *Question* on the inherent necessity of both the substance of the sphere and its motion is due to a realization of some inadequacy in Alexander's distinction between the meaning of the term "force" when applied to a mover, and the meaning of that term when applied to what undergoes motion. (See the passages cited in n. 43, below.)

39. **Distinguished:** see above, *Question* IV, n. 30.

40. See, e.g., *Question* VII, paragraph 48; other passages are cited in the notes *ad loc.* See also, H. Davidson, "The Principle That a Finite Body Can Contain only Finite Power," *Studies in Jewish Religious and Intellectual History*, ed. S. Stein and R. Loewe (University, Alabama; 1979), pp. 75-85.

41. See Wolfson, *Crescas*, pp. 654 ff.

42. I.e., with respect to any kind of infinite intensity.

43. See *Question* VII, paragraph 48; *LP* VIII, Comm. 79, 426 M-427 C; *LM* XII, Comm. 41, 324 C-H (Arabic: pp. 1629, 1-1631, 9); *IDC* I, x, 2, 8, fol. 178a, 8-180a, 24 (Latin: 293 K-295 B); *LDC* II, Comm. 71, 145 I-146 G; *DSO* III, pp. 102-103 (Hebrew: p. 38, 31-37; Latin: 9 BC).

In his *Proofs*, ch. x, Davidson has subjected Averroes' critique of Avicenna to very stringent scrutiny, and has made considerable use of this *Question* for that purpose. His discussion deserves the most careful study--unflattering as his conclusions are to Averroes. I do not wish to enter into dispute here, but simply to raise a consideration I think Professor Davidson has insufficiently stressed.

There are certainly problems with the way Averroes proceeds in this treatise (and elsewhere). As Professor Davidson notes, he is apparently unwilling (or unable) to engage Avicenna on Avicenna's ground. But to do so, would require concessions that Averroes is not prepared to make.

In the short compass of this *Question*, Averroes does not even attempt to lay out all the sources of his disagreement with Avicenna, even when they are directly relevant to his topic. For example, we have only the barest mention of the dispute over Avicenna's split of existence from essence. He does not even lay out, to the extent he does in *TAT*, what the problems are with regard to Avicenna's use of the term "possible." As we have said, what Averroes does do is to continue to use terms, such as possibility and necessity, in the sense he thought was truly Aristotelian, rather than in the sense Avicenna intended. It is not clear to me that this is due to ignorance; Averroes does much the same kind of thing when he deals with the Kalâm (although it must be admitted that there is some doubt as to his knowledge of this body of literature, as well). But this does appear to be his general method of dealing with opponents whose premises he finds unacceptable. And Avicenna is not acceptable to the mature Ibn Rushd, any more than the Kalâm, or John Philoponus, is. The nature of the universe in Avicenna's system, eternal and necessary though it is, cannot satisfy Averroes, for his opponents are not simply those who, on religious grounds, uphold the doctrine of *creatio ex nihilo*. He stands in opposition to all those who seek an ultimate cause for existence, and he identifies his opponents with the disciples of Plato. He, like Aristotle, seeks only an ultimate cause for motion. This lands him with the same kinds of problems that Aristotle faced, plus the necessity for answering criticisms of Aristotle that had accumulated over the centuries. Averroes is not always successful in his quest for solutions, but he cannot abandon Aristotle. In the search for an ultimate cause of motion, there was no other role model. There really is no common ground between Averroes and his opponents; they are not talking about the same thing.

Perhaps the most troubling aspect of the present treatise is the fact that Averroes never makes it clear why a motion necessary in and of itself should require a mover, or in what way the mover of such a motion can be said to be a cause. This is similar to the question with which *Question* V leaves us: if there are two eternal existences (and eternal means necessary), what is the nature of the causal relation between them? Averroes does discuss that problem in other writings, but nothing in his solutions would be compatible with the view expressed here that the motion of the sphere contains no possibility at all.

44. **Aristotelians**: lit., the party of Aristotle of the Peripatetics. **Of the Peripatetics** (*mê-ha-maśśâ'iyîm*) is probably a gloss incorporated into the text. It may, however, be a corruption of "of the commentators" (*mê-ha-mepâreshîm*); cf. *EP* I, p. 12, 5-6 (Arabic): "... and indeed [this] is the literal meaning of the discourse of Aristotle and those who followed him of the commentators...".

45. **Reputable**: *mequbbâlîm*; perhaps the equivalent of *endoxoi*. See Walzer, *Greek into Arabic*, p. 95.

BIBLIOGRAPHY

1. Works by Averroes

1.1. EPITOMES

De Anima. Arabic: *Kitâb al-Nafs*, in *Rasâ'il Ibn Rushd*, Hyderabad, 1947.
De Caelo. Arabic: *Kitâb al-Sama' wa al-'Alam*, in *Rasâ'il Ibn Rushd*, Hyderabad, 1947.
De Generatione Animalium. Latin: in *Aristotelis omnia quae extant Opera*, vol. VI, 2, Venetiis, 1574; reprinted Frankfurt a.M., 1962.
De Generatione et Corruptione. Arabic: *Kitâb al-Kawn wa al-Fasâd*, in *Rasâ'il*; Hebrew: *Qiṣṣûr Sêper ha-Hawâyâ wa ha-Hefsêd le-Abu al-Walîd Ibn Rushd*, ed. S. Kurland, *CCAA*, vol. IV, 1-2, Cambridge, Mass., 1958; English: S. Kurland, *CCAA*, vol. IV, 1-2, Cambridge, Mass., 1958.
Metaphysics. Arabic: *Kitâb mâ ba'ad al-Ṭabî'a* in *Rasâ'il*; Latin: in *Opera*, vol. VIII; German: S. van den Bergh, *Die Epitome der Metaphysik des Averroes*, Leiden, 1924.
Organon. Arabic: C.E. Butterworth, *Averroes' Three Short Commentaries on Aristotle's "Topics," "Rhetoric," and "Poetics"* [with English translation], Albany, 1977; Hebrew: *Kol Meleket Higgâyôn*, Riva di Trento, 1560; Latin: in *Opera*, vol. I, 2b and vol. II.
Parva Naturalia. Arabic: *Talkhîs Kitâb al-Ḥiss wa al-Maḥsûs*, ed. H. Blumberg, *CCAA*, vol. VII, Cambridge, Mass., 1972; Hebrew: *Qiṣṣûr Sêper ha-Ḥûš we ha-Mûḥaš*, ed. H. Blumberg, *CCAA*, vol. VII, Cambridge, Mass., 1954; Latin: ed. E. Shields, *CCAA*, vol. VII, Cambridge, Mass., 1949, and in *Opera*, vol. VI, 2; English: H. Blumberg, *CCAA*, vol. VII, Cambridge, Mass., 1961.
Physics. Arabic: *Kitâb al-Samâ' al-Ṭabî'î*, ed. J. Puig, *CCAA*, vol. XX, Madrid, 1983; Hebrew: *Qiṣṣûrê Ibn Rushd 'al Šêma' Ṭib'îyî le-Aristoteles*, Riva di Trento, 1560.

1.2. INTERMEDIATE COMMENTARIES

Categories. Arabic: *Talkhîç Kitâb al-Maqoulat*, ed. M. Bouyges, Beyrouth, 1932, and *Talkhîs Kitâb al-Maqûlât*, ed. M. Kassem, C.E. Butterworth, and A.A. Haridi, *CCAA*, vol. 1, a (2), Cairo, 1980; and A.A. Haridi, *CCAA*, vol. I, a (2), Cairo, 1980; Hebrew: *Ha-Bê'ûr ha-'emṣâ'î shel Ibn Rushd 'al Sêper ha-Ma'amârôt*, ed. H. Davidson, *CCAA*, vol. Ia, 1-2, Cambridge, Mass., 1969; Latin: in *Opera*, vol. I, 1.
De Caelo. Hebrew: Hebrew MS. 42, Harvard; Latin: in *Opera*, vol. V.
De Generatione et Corruptione. Hebrew: *Ha-bê'ûr ha-'emṣâ'î 'al Sêper ha-Hawâyâ we ha-Hefsêd*, ed. S. Kurland, *CCAA*, vol IV, 1-2, Cambridge, Mass., 1958; Latin: ed. F.H. Fobes, *CCAA*, vol. IV, 1, Cambridge, Mass., 1956, and in *Opera*, vol. V; English: S. Kurland, *CCAA*, vol. IV, 1-2, Cambridge, Mass., 1958.
De Interpretatione. Arabic: *Talkhîs Kitâb al-'ibâra*, ed. M. Kassem, with C.E. Butterworth and A.A. Haridi, *CCAA*, Cairo, 1981.

De Sophisticis Elenchis. Latin: in *Opera*, vol. I, 3.
Isagoge. Hebrew: *Ha-bê'ûr ha-'emṣâ'î 'al Sêper ha-Mâbô'*, ed. H. Davidson, *CCAA*, vol. Ia, 1-2, Cambridge, Mass., 1969; Latin: in *Opera*, vol. I, 1.
Physics. Hebrew: Hebrew MS. 42, Harvard; Latin [Bks. I-III]: in *Opera*, vol. IV.
Prior Analytics. Latin: in *Opera*, vol. I, 1.
Posterior Analytics. Arabic: *Talkhîs Kitâb al-Burhân*, ed. M. Kassem, C.E. Butterworth, and A.A. Haridi, *CCAA*, vol 1, a (5), Cairo, 1982.
Republic. Hebrew: ed. E.I.J. Rosenthal [with English translation], *Averroes' Commentary on Plato's Republic*, Cambridge, 1966; English: R. Lerner, *Averroes on Plato's Republic*, Ithaca, 1974.
Rhetoric. Latin: in *Opera*, vol. II.
Topics. Arabic: *Talkhîs Kitâb al-Djadal*, ed. C.E. Butterworth and A.A. Haridi, *CCAA*, vol. 1, a (6), Cairo, 1979; Latin: in *Opera*, vol. I, 3.

1.3. LONG COMMENTARIES

De Anima. Latin: ed. F. Crawford, *CCAA*, vol. VI, 1, Cambridge, Mass., 1953; and in *Opera*, vol. VI, 1.
De Caelo. Latin: in *Opera*, vol. V.
Metaphysics. Arabic: *Tafsîr mâ ba'ad aṭ-Ṭabî'at*, 3 vols., ed. M. Bouyges, Beyrouth, 1938-1948; Latin: in *Opera*, vol. VIII; Hebrew [a portion of Bk. XII, Comm. 18]: Munich MS. 36, fols. 221v-222v; English [Bk. XII]: C. Genequand, *Ibn Rushd's Metaphysics*, Leiden, 1984.
Physics. Latin: in *Opera*, vol. IV.
Posterior Analytics. Latin: in *Opera*, vol. I, 2a.

1.4. OTHER WORKS

Colliget [Kûllîyyût]. Latin: in *Opera*, vol. X.
De Spermate. Latin: *Aristotelis omnia quae extant Opera*, vol. XI, Venetiis, 1560.
Kitâb Faṣl al-Maqâl. Arabic: ed. G. Hourani, Leiden, 1959; English: G. Hourani, *Averroes on the Harmony of Religion and Philosophy*, "E.J.W. Gibb Memorial" Series, New Series, XXI, London, 1961.
Kitâb al-Kašf 'an Manâhidj al-'Adilla. Arabic: in *Falsafa Ibn Rushd*, ed. M.A. Sabih, 2nd ed., Cairo, 1935; Arabic text with German translation: in M.J. Müller, *Die Philosophie und Theologie von Averroes*, München, 1875.
Quaesita varia in Logica. Latin: in *Opera*, vol. I, 2a.
Sermo de Substantia Orbis. Hebrew [with English translation]: A. Hyman, *Ma'amâr be-'eṣem ha-galgal [Averroes on the Substance of the Celestial Sphere]*, *CCAA*, Cambridge, Mass. and Jerusalem, 1986.
Tahâfut at-Tahâfut. Arabic: ed. M. Bouyges, Beyrouth, 1930; Latin: in *Opera*, vol. IX, and ed. B. Zedler, *Averroes' Destructio Destructionum Philosophiae Algazelis*, Milwaukee, 1961; English: S. van den Bergh, *Averroes' Tahafut al-Tahafut (The Incoherence of the In-*

coherence, "E.J.W. Gibb Memorial" Series, New Series, vols. XVIII-XIX, London, 1954.

2. Works by Aristotle

Categories. Greek: ed. L. Minio-Paluello, Oxford, 1949; English: E.M. Edgehill, Oxford, 1928 [in *The Works of Aristotle Translated into English*, ed. W.D. Ross, Oxford, 1908-1931]; Arabic: *Kitâb al-Maqûlât* in *Manṭiq Arisṭû*, ed. A. Badawi, vol. I, Cairo, 1948.

De Anima. Greek: ed. W.D. Ross, Oxford, 1961; English: J.A. Smith, Oxford, 1931 [in *Works*].

De Caelo. Greek: ed. D.J. Allan, Oxford, 1936; English: J.L. Stocks, Oxford, 1922 [in *Works*]. On the Arabic translation, see, in section 3 of this bibliography, **Endress**.

De Generatione Animalium. Greek: ed. H.J. Drossaart Lulofs, Oxford, 1965; English: A. Platt, Oxford, 1910 [in *Works*]; Arabic: *Fî Kawn al-Ḥayawân*, ed. J. Brugman and H.J. Drossaart Lulofs, Publication of the 'De Goeje Fund,' Nr. XXIII, Leiden, 1971.

De Generatione et Corruptione. Greek: ed. H. Joachim, Oxford, 1922; English: H. Joachim, Oxford, 1922 [in *Works*].

De Interpretatione. Greek: ed. L. Minio-Paluello, Oxford, 1949; English: E.M. Edgehill, Oxford 1928 [in *Works*]; Arabic: in *Alfarabi's Commentary on Aristotle's De Interpretatione*, ed. W. Kutsch and S. Marrow, 2nd ed., Beyrouth, 1971.

De Motu Animalium. Greek: ed. L. Torraca, *Collana de Studi Greci*, vol. XXX, Napoli, 1958; English: A.S.L. Farquharson, Oxford, 1912 [in *Works*].

De Sophisticis Elenchis. Greek: ed. W.D. Ross, Oxford, 1958; English: W.A. Pickard-Cambridge, Oxford, 1928 [in *Works*]; Arabic: *Kitâb al-Sûfisṭîqâ*, ed. A. Badawi, in *Manṭiq*, vol. III, Cairo, 1952.

Metaphysics. Greek: ed. W.D. Ross, 2 vols., Oxford, 1924; English: W.D. Ross, Oxford, 1928 [in *Works*]; Arabic: printed together with Averroes' Long Commentary in *Tafsîr mâ ba'ad aṭ-Ṭabî'at*.

Meteorologica. Greek: ed. F.H. Fobes, Cambridge, Mass., 1918; English: E.W. Webster, Oxford, 1923 [in *Works*].

Physics. Greek: ed. W.D. Ross, Oxford, 1936; English: R.P. Hardie and R.K. Gaye, Oxford, 1930 [in *Works*]; *Arabic:* Al-Ṭabî'a, *ed.* A. Badawi, 2 vols., Cairo, 1964-65.

Posterior Analytics. Greek: *Aristotle's Prior and Posterior Analytics*, ed. W.D. Ross, Oxford, 1949; English: G. Mure, Oxford, 1928 [in *Works*]; Arabic: *Kitâb al-Taḥlîlât al-Thânîya* (*Kitâb al-Birhân*), ed. A. Badawi, *Manṭiq*, vol. II, Cairo, 1949.

Prior Analytics. Greek: ed. W.D. Ross, Oxford, 1949; English: A. Jenkinson, Oxford, 1928 [in *Works*]; Arabic: *Kitâb al-Taḥlîlât al-'awlî*, ed. A. Badawi, *Manṭiq*, vol. I, Cairo, 1948.

Rhetoric. Greek: ed. W.D. Ross, Oxford, 1959; English: W. Rhys Roberts, Oxford, repr. 1946 [in *Works*].

Topics. Greek: ed. W.D. Ross, Oxford, 1958; English: W.A. Pickard-Cambridge, Oxford, 1928 [in *Works*]; Arabic: *Kitâb al-Tûpîqâ*, ed. A. Badawi, *Manṭiq*, vol. III, Cairo, 1952.

3. Works by Other Authors

Abraham Shalom, *Neveh Šâlôm*, Venice, 1575.
Abravanel, Isaac, *Mip'âlôt 'Elôhîm, Lemberg, 1863.*
-- *Šâmayîm Ḥadâšîm*, Rodelheim, 1828.
Abu'l-Faraj ibn al-Ṭayyib, *Commentary to Aristotle's Physics*, printed together with the Arabic version of *Physics*, ed. A. Badawi.
Alexander Aphrodisiensis, *Alexandri Aphrodisiensis Quaestionum Naturalium et moralium ad Aristotelis Philosophiam illustrandum*, ed. L. Spengel, Monachii, 1842.
------ *De Principiis* (*Fî mabâdî al-kull*), in *Arisṭû 'inda-l-'Arâb*, ed. A. Badawi, Cairo, 1947.
------ *In Aristotelis Analyticorum Priorum Librum I Commentarium*, ed. M. Wallies, *CAG* II, 1, 1883.
------ "Refutation of Galen," *see below*, Pines.
------ *Treatise on Time* (*Maqâla fî al-zamân*), in A. Badawi, *Commentaires sur Aristote perdus en grec et autres épitres*, Beyrouth, 1971. *See also* Pines.
Al-Farabi, *Ārâ' Ahl al-Madîna al-Fâḍila* (*Der Musterstaat*), ed. F. Dieterici, repr. Leiden, 1964. *See also* Walzer.
------ "Die Harmonie zwischen Plato und Aristoteles," *Alfarabi's philosophische Abhandlungen*, trans. F. Dieterici, Leiden, 1892; Arabic text: *Alfarabi's philosophische Abhandlungen*, ed. F. Dieterici, Leiden, 1890. *See also* Mahdi.
Alonso, M.A., *Teología de Averroës (estudios y documentos)*, Madrid-Granada, 1947.
Asclepius, *Asclepii in Metaphysicorum, Libros A-Z Commentaria*, ed. M. Hayduck, *CAG* VI, 2, 1888.
Ammonius, *Ammonii in Aristotelis Analyticorum Priorum Librum I Commentarium*, ed. M. Wallies, *CAG* IV, 6, 1899.
Asin Palacios, M., *Abenhazam de Córdoba*, 5 vols., Madrid, 1927-32.
Assemanus, Stephanus Evodius et Josephus Simonius, *Bibliothecae Apostolicae Vaticanae codicum manuscriptorum Catalogus*, vol. I, i, Rome, 1756.
Avicenna, *Najât*, Rome, 1593.
------ *Šifâ'*, *Avicenne: La Métaphysique du Shifâ'*, trans. G. Anawati, 2 vols., Paris, 1978/1985.
See also Horten, Morewedge.
Badawi, A., ed., *Arisṭû 'ind al-'Arab*, Cairo, 1947.
------ *Commentaires sur Aristote perdus en grec et autres épitres*, Beyrouth, 1971.
Barnes, J., "Aristotle's Theory of Demonstration," in *Articles on Aristotle* 1, London, 1975, pp. 65-87.
Birkenmajer, A., "Eine wiedergefundene Übersetzung Gerhards von Cremona," *Beiträge zur Geschichte der Philosophie und Theologie des Mittelalters*, Supplement Band 3, Festschrift Grabmann, 1935, pp. 472-481.
Bochenski, I.M., *A History of Formal Logic*, trans. and ed. by Ivo Thomas, New York, 1970.
Booth, E., *Aristotelian Aporetic Ontology in Islamic and Christian Thinkers*, *Cambridge Studies in Mediaeval Life and Thought*, 3rd series, vol. 20, Cambridge, 1983.

Brown, H.V.B., "Avicenna and the Christian Philosophers in Bagdad," in *Islamic Philosophy and the Classical Tradition*, ed. S. Sterns, et al., Oxford, 1972, pp. 35-48.

Carra de Vaux, B., *Gazali*, Paris, 1902, repr. Amsterdam, 1974.

Casiri, M., *Bibliotheca Arabico-Hispana Escurialensis*, 2. vols., Madrid, 1740.

Corbin, H., *Avicenna and the Visionary Recital*, Bollingen Series LXVI, New York, 1960.

Cleary, J., *Aristotle on the Many Senses of Priority*, *Journal of the History of Philosophy Monograph Series*, Southern Illinois University Press, Carbondale and Edwardsville, 1988.

Crescas, Hasdai, *Ôr 'Adônâi*, printed together with the text of Maimonides, *Môrê Nebûkîm*, repr. New York, 1946. *See also* Wolfson.

Davidson, H.A., "John Philoponus as a Source of Medieval Islamic and Jewish Proofs of Creation," *JAOS* 89, 1969, pp. 357-391.

------ *The Philosophy of Abraham Shalom*, University of California Publications, Near Eastern Studies, V, Berkeley and Los Angeles, 1964.

------ "The Principle that a Finite Body Can Contain Only Finite Power," in *Studies in Jewish Religious and Intellectual History*, ed. S. Stein and R. Loewe, University, Alabama, 1979, pp. 75-92.

------ *Proofs for Eternity, Creation and the Existence of God in Medieval Islamic and Jewish Philosophy*, New York/Oxford, 1987.

Derenbourg, H., *Les manuscrits arabes de l'Escurial*, vol. I, Paris, 1884.

Drossaart Lulofs, H., *Nicolaus Damascenus on the Philosophy of Aristotle*, Leiden, 1965.

Duhem, P., *Le système du monde*, 5 vols., Paris, 1913-17.

Duran, Simon, *Qešet û-Mâgên*, ed. M. Steinschneider, *'Ôṣâr Ṭôb* VII, 1881.

Efros, I., *Maimonides' Treatise on Logic*. *Proceedings, American Academy for Jewish Research* VIII, 1938.

Elias Cretensis [Elijah del Medigo], *Heliae Cretensis Phylosophi, et Medici, in dictis Averrois super libros physicorum clarissimae adnotationes*, Venetiis, 1551.

Endress, G., *Die arabischen Übersetzungen von Aristoteles' Schrift De Caelo*, Frankfurt a.M., 1966.

Evrard, E., "Les convictions religieuses de Jean Philopon et la date de son Commentaire aux 'Météorologiques'," *Academie royale de Belgique, Bulletin de la Classe des Lettres et des Sciences Morales et Politiques*, 5e série-Tome XXXIX, 1953, pp. 299-357.

Fleischer, H., and Delitzsch, F., *Catalogus librorum manuscriptorum qui in Bibliotheca senatoria civitatis lipsiensis asservantur* (Codices orientalium linguarum), Grimae, 1838.

Frank, R., "Bodies and Atoms: The Ash'arite Analysis," in *Islamic Theology and Philosophy: Studies in Honor of George F. Hourani*, ed. M. Marmura, Albany, 1984, pp. 39-53.

Galen, *On the Natural Faculties*, trans. A. Brock, Loeb Classical Library, London and New York, 1916. *See also* Kieffer.

Gersonides, *Milḥâmôt ha-Šêm*, Leipzig, 1866.

Goldstein, H. Tunik, "Averroes on the Structure and Function of *Physics* VII, 1," *Harry Austryn Jubilee Volume* I, American Academy for Jewish Research, Jerusalem, 1965.

------ "*Dator Formarum*: Ibn Rushd, Levi b. Gerson, and Moses b. Joshua of Narbonne," *Islamic Thought and Culture: Papers Presented to the Islamic Studies Group of the American Academy of Religion*, International Institute of Islamic Thought, 1982, pp. 107-121.
------ "New Hebrew Manuscript Sources for Averroean Texts," *JNES* 38, 1979, pp. 29-31.
Grosseteste, Robert, *Commentarius in VIII Libros Physicorum Aristotelis*, ed. R.C. Dales, Boulder, Colorado, 1963.
Gutas, D., *Avicenna and the Aristotelian Tradition*, Leiden, 1988.
Gyekye, K., "The Term Isti<u>th</u>nâ' in Arabic Logic," *JAOS* 92, 1972, pp. 88-92.
Haarbrücker, T., *Asch-schahrastani's Religionspartheien und Philosophenschulen*, 2 vols., Halle, 1850-51.
Hintikka, J., "Necessity, Universality, and Time in Aristotle," in *Articles on Aristotle* 3, London, 1979, pp. 108-124.
------ "Aristotelian Infinity," in *Articles on Aristotle* 3, London, 1979, pp. 125-139.
Horten, M., *Die Metaphysik Avicennas*, Halle, 1907-09.
Hourani, G., ed., *Essays on Islamic Philosophy and Science*,, Albany, 1975.
Hugonnard Roche, H., "L'Épitomé du *De Caelo* d'Aristote par Averroès: Questions de méthode et de doctrine," *AHDLMA* 51, 1984, pp. 7-39.
------ "Méthodes d'argumentation et philosophie naturelle chez Averroès, *Orientalische Kultur und Europäisches Mittelalter* (*Miscellanea Mediaevalia* 17), Berlin, 1985, pp. 240-253.
Hyman, A., "The Composition and Transmission of Averroes' *'Ma'amar be-'Eṣem ha-Galgal*,'" *Studies and Essays in Honor of Abraham A. Neuman*, Leiden, 1962, pp. 299-307.
Isaac Albalag, *Tiqqûn ha-Dê'ôt*, ed. O. Schorr, *He-Ḥâlûṣ* VI, 1861, pp. 85-94. *See also* Vajda.
Ivry, A., "Destiny Revisited: Avicenna's Concept of Determinism," in *Islamic Theology and Philosophy: Studies in Honor of George F. Hourani*, ed. M. Marmura, Albany, 1984, pp. 160-171.
Jaeger, W., *Aristotle: Fundamentals of the History of his Development*, tr. with the author's corrections and additions by R. Robinson, Oxford, 1934.
Jalbert, G., "La necessité et la contingence chez Aristote et Averroès," *Revue de l'Université d'Ottawa XXX, 1960, pp. 21*-36**
John Philoponus, *Ioannis Philoponi in Aristotelis Physicorum Libros Tres Priores Commentaria*, ed. H. Vitelli, *CAG* XVI, 1887.
------ *Ioannis Philoponi in Aristotelis Physicorum Libros Quinque Posteriores Commentaria*, ed. H. Vitelli, *CAG* XVII, 1888.
See also Kraemer, Mahdi, Pines.
Judah ha-Levi, *Kuzari*, Arabic: *Kitâb al-Radd wa-'l-Dalîl fî 'l-Dîn al-<u>Dh</u>alîl*, ed. D. Baneth, Jerusalem, 1977; *Kitâb al-Hu<u>dj</u>d<u>j</u>a wa al-Dalîl fî nasr al-Dîn al-<u>Dh</u>alîl*, ed. H. Hirschfeld, Leipzig, 1887. Hebrew translation by Yehudah Ibn Tibbon, repr. New York, 1946. English translation from the Arabic by H. Hirschfeld, London, 1906.
Kaufmann, D., *Geschichte der Attributenlehre in der jüdischen Religionsphilosophie des Mittelalters von Saadja bis Maimuni*, Gotha, 1877.

Kholeif, F., *A Study on Fakhr al-Dîn al-Râzî and his controversies in Transoxiana*, Beyrouth, 1966.
Kieffer, J.S., *Galen's Institutio Logica*, Baltimore, 1964.
Kogan, B., "Eternity and Origination: Averroes' Discourse on the Manner of the World's Existence," in *Islamic Theology and Philosophy: Studies in Honor of George F. Hourani*, ed. M. Marmura, Albany, 1984, pp. 203-35.
------ *Averroes and the Metaphysics of Causation*," Albany, 1984.
Kraemer, J., "A Lost Passage from Philoponus' *Contra Aristotelem*, in Arabic Translation," *JAOS* 85, 1965, pp. 318-327.
Lejewski, C., "On Prosleptic Syllogisms," *Notre Dame Journal of Formal Logic* II, 1961, pp. 158-76.
Lewin, B., "La notion de muḥdath dans le Kalam et dans la philosophie," *Orientalia Suecana* 3, 1954, pp. 84-93.
Madkour, I., *L'Organon d'Aristote dans le monde arabe*, Paris, 1934.
Mahdi, M., "Alfarabi against Philoponus," *JNES* 26, 1967, pp. 233-260.
------ *Alfarabi's Philosophy of Plato and Aristotle*, [trans., with an introduction], Glencoe, Ill., 1962.
------ "The Arabic Text of Alfarabi's Against John the Grammarian," *Medieval and Middle Eastern Studies in Honor of A.S. Atiya*, ed. S. Hanna, Leiden, 1972, pp. 268-84.
Maimonides, *Millôt ha-Higgâyôn*, ed. L. Roth and D. Baneth, Jerusalem, 1935. Arabic: *Maqâla fî ṣinâʿa al-manṭiq*, ed. I. Efros, *Maimonides' Treatise on Logic, Proceedings, American Academy for Jewish Research* VIII, 1938.
------ *Môrê Nebûkîm*, repr. New York, 1946. Arabic: *Dalâla al-Ḥâ'irîn*, ed. J. Junowitz, Jerusalem, 1929. English translations, from the Arabic, by M. Friedlander, New York, reprinted from the 1881 edition, n.d.; and S. Pines, Chicago, 1963.
Marmura, M., ed., *Islamic Theology and Philosophy: Studies in Honor of George F. Hourani*, Albany, 1984.
Morewedge, P., *The Metaphysica of Avicenna*, Persian Heritage Series, No. 13, Columbia University Press, New York, 1973.
Munk, S., *Mélanges de philosophie juive et arabe*, Paris, 1927.
Nallino, C.A., "Filosofia 'orientale' od 'illuminativa' d' Avicenna?", *Rivista degli Studi Orientali* X, 1923-25, pp. 433-467.
Narboni, *Commentary* to *Môrê Nebûkîm*, printed together with that text, repr. New York, 1946.
Nicolaus Damascenus, *see* Drossaart Lulofs.
Olympiodorus, *Prolegomena et in Categorias Commentarium*, ed. A. Busse, *CAG* XII, 1, 1902.
Pasini, J., *Codices Manuscripti Bibliothecae R. Taurinensis*, vol. I, Taurini, 1749.
Peters, F.E., *Aristoteles Arabus, Monographs on Mediterranean Antiquity* II, Leiden, 1968.
------ *Aristotle and the Arabs: The Aristotelian Tradition in Islam*, New York, 1968.
Pines, S., "An Arabic Summary of a Lost Work of John Philoponus," *Israel Oriental Studies* II, 1972, pp. 320-352.
------ "*Omne quod movetur necesse est ab aliquo moveri*: A Refutation of Galen by Alexander of Aphrodisias and the Theory of Motion," *Isis* 52, 1961, pp. 21-54.

------ "La 'Philosophie orientale' d'Avicenne et sa polémique contre les Bagdadiens," *AHDLMA* XIX, 1952, pp. 5-37.
See also Sambursky.

Plato, *Laws*.

------ *Phaedrus*.

Plutarch, *Lives*, Dryden translation, New York, n.d.

Prantl, Carl von, *Geschichte der Logik im Abendlande*, 4 vols. in 2, repr. Leipzig, 1927.

Proclus, *Procli Diadochi in Platonis Timaeum Commentaria*, ed. E. Diehl, Leipzig, 1903.

Quran, trans. M. Rodwell, London, 1909.

Rahman, F., "The Eternity of the World and the Heavenly Bodies in Post-Avicennan Philosophy," in *Essays on Islamic Philosophy and Science*, ed. G. Hourani, Albany, 1975, pp. 222-237.

Renan, E., *Averroès et l'averroïsme*, Paris, 1861.

Rosenberg, S., "The Hebrew Translations of Averroes' *Questions in Physics* and the Commentary of Moses Narboni," *Qiryat Sêper* 57, 1982, pp. 715-24 [in Hebrew].

Rosenthal, F., "A Commentator of Aristotle," in *Islamic Philosophy and the Classical Tradition*, ed. S. Stern et al., Oxford, 1972, pp. 337-49.

------ "From Arabic Books and Manuscripts V: A One-Volume Library of Arabic Philosophical and Scientific Texts in Istanbul," *JAOS* 75, 1955, pp. 14-23.

Sambursky, S., and Pines, S., *The Concept of Time in Later Neoplatonism*, Jerusalem, 1971.

Schmitt, C., "Renaissance Averroism Studied through the Venetian Editions of Aristotle and Averroes (with Particular Reference to the Giunta Edition of 1550-1552)," in *The Aristotelian Tradition and Renaissance Universities*, No. VIII, London, 1984.

Schmoelders, A., *Essai sur les écoles philosophiques chez les arabes*, Paris, 1842.

Schwab, M., "Manuscrits du supplément hébreu de la Bibliothèque Nationale," *Revue des études juives* XXXVII, 1898, pp. 127-36.

Shem Tob [b. Joseph b. Shem Tob b. Shem Tob], *Commentary* to *Môrê Nebûkim*, printed together with the text of that work, repr. New York, 1946.

Simplicius, *Simplicii in Aristotelis Categorias Commentarium*, ed. C. Kalbfleisch, *CAG* VIII, 1907.

------ *Simplicii in Aristotelis De Caelo Commentaria*, ed. I.L. Heiberg, *CAG* VII, 1894.

------ *Simplicii in Aristotelis Physicorum Libros Quattuor Posteriores Commentaria*, ed. H. Diels, *CAG* X, 1895.

Sorabji, R., "Infinity and the Creation," in *Philoponus and the Rejection of Aristotelian Science*, London, 1987, pp. 164-178.

------ *Necessity, Cause, and Blame*, Ithaca, N.Y., 1980.

------ ed., *Philoponus and the Rejection of Aristotelian Science*, London, 1987.

------ *Time, Creation and the Continuum: theories in antiquity and the early middle ages*, London, 1983.

Stein, S., and Loewe, R., eds., *Studies in Jewish Religious and Intellectual History Presented to Alexander Altmann on the Occasion of His Seventieth Birthday*, University, Alabama, 1979.
Steinschneider, M., "Alfarabi," *Mémoires de l'academie impériale des sciences de St.-Petersbourg*, VIIe Série, XIII, No. 4, 1896, pp. 1-268.
------ "Elijah del Medigo," *Hebraeische Bibliographie* XXI, 1881/82, pp. 60-71.
------ *Die Handschriften-verzeichnisse der Königliche Bibliothek zu Berlin*, Zweiter Band. *Verzeichniss der Hebraeischen Handschriften*, Berlin, 1878.
------ *Die Hebraeischen Handschriften der K. Hof-und Staatsbibliothek in Muenchen*, Muenchen, 1875; 2nd ed., 1895.
------ "Mündlich oder schriftlich?", *Hebraeische Bibliographie* XVII, 1870, pp. 17-19.
------ "Anzeigen (Philosophie, Fortsetz.)," *Hebraeische Bibliographie* X, 1870, pp. 53-59.
------ *Die Hebräischen Übersetzungen des Mittelalters und die Juden als Dolmetscher*, Berlin, 1893.
Stephanus, *In De Interpretatione Commentarium*, ed. M. Hayduck, *CAG* XVIII, 3, 1895.
Stern, S.M., "Ibn al-Samḥ," *Journal of the Royal Asiatic Society*, April, 1956, pp. 31-44.
------ et al., eds., *Islamic Philosophy and the Classical Tradition: Essays presented by his friends and pupils to Richard Walzer on his seventieth birthday*, *Oriental Studies* 5, Oxford, 1973.
Tannery, P., "Sur la période finale de la philosophie grecque," *Revue philosophique* XLII, 1896, pp. 266-287.
Teicher, J.L., "Averroès inconnu," *Actes du XXe Congrès International des Orientalistes*, 1938, pp. 337-38.
------ "Avicenna's Place in Arabic Philosophy," in *Avicenna: Scientist and Philosopher*, A Millenary Symposium, ed. G.M. Wickens, London, 1952.
Themistius, *Themistii in Libros Aristotelis De Caelo Paraphrasis*, Hebraice et Latine, ed. S. Landauer, *CAG* V, 4, 1902.
------ *Themistii in Aristotelis Physica Paraphrasis*, ed. H. Schenkl, *CAG* V, 2, 1900.
Vajda, G., "Les deux versions hébraïques de la dissertation d'Averroès sur la science divine," *Revue des études juives*, n.s. 13, 1954, pp. 63-66.
------ *Isaac Albalag, Études de philosophie mediévale*, 49, Paris, 1960.
------ *Recherches sur la philosophie et la kabbale dans la pensée juive du Moyen Âge*, Paris, 1962.
Walzer, R., *Alfarabi on the Perfect State*, Oxford, 1985.
------ *Greek into Arabic: Essays on Islamic Philosophy*, *Oriental Studies* I, Cambridge, Mass, 1962.
Wieland, W., "Die Ewigkeit der Welt (der Streit zwischen Joannes Philoponus und Simplicius)," in D. Heinrich et al., eds., *Die Gegenwart der Griechen im neueren Denken, Festschrift H.-G. Gadamer*, Tübingen, 1960, pp. 291-316.
Wildberg, C., *Philoponus. Against Aristotle on the Eternity of the World*, London, 1987.

Wolff, M., "Philoponus and the Rise of Preclassical Dynamics," in R. Sorabji, ed., *Philoponus and the Rejection of Aristotelian Science*, London, 1987.

Wolfson, H.A., "Averroes' Lost Treatise on the Prime Mover," *Hebrew Union College Annual* XIII, Part One, 1950-51, pp. 683-710.

------ *Crescas' Critique of Aristotle*, Cambridge, Mass., 1929.

------ "Hallevi and Maimonides on Design, Chance and Necessity," *Proceedings of the American Academy for Jewish Research* XI, 1941, pp. 105-63.

------ "The Kalam Arguments for Creation in Saadia, Averroes, Maimonides and St. Thomas," *Saadia Anniversary Volume* II, *American Academy for Jewish Research*, New York, 1943, pp. 197-245.

------ "Note on Crescas' Definition of Time," repr. from *Jewish Quarterly Review*, N.S. X, 1, Philadelphia, 1919, 17 pp.

------ "Notes on Proof of the Existence of God in Jewish Philosophy," *Hebrew Union College Annual* I, 1924, pp. 575-96.

------ *The Philosophy of the Kalam*, Cambridge, Mass. and London, 1976.

------ *The Philosophy of Spinoza*, Cambridge, Mass., 1948.

------ "The Platonic, Aristotelian, and Stoic Theories of Creation in Hallevi and Maimonides," *Essays in honour of the Very Rev. Dr. J.H. Hertz*, London, 1942, pp. 427-42.

------ "The Plurality of Immovable Movers in Aristotle, Averroes, and St. Thomas," *Harvard Studies in Classical Philology* LXIII, 1958, pp. 233-53.

------ "The Problem of the Souls of the Spheres from the Byzantine Commentaries to Aristotle through the Arabs and St. Thomas to Kepler," *Dumbarton Oaks Papers* 16, 1962, pp. 67-93.

------ *Repercussions of the Kalam in Jewish Philosophy*, Cambridge, Mass., and London, 1979.

------ "Revised Plan for the Publication of a *Corpus Commentariorum Averrois in Aristotelem*," *Speculum* 37, 1963, pp. 88-104.

------ "The Twice Revealed Averroes," *Speculum* 36, 1961, pp. 373-393.

Worms, M., *Die Lehre von der Anfanglosigkeit der Welt bei den mittelalterlichen arabischen Philosophen des Orients und ihre Bekämpfung durch die arabischen Theologen (Mutakallimûn)*, *Beiträge zur Geschichte der Philosophie des Mittelalters* III, iv, Münster, 1900.

Yaḥya ibn ʻAdi, *Commentary* to Aristotle's *Physics*, printed together with the Arabic text of *Physics*, ed. A. Badawi, Cairo, 1964-65.

Zimara, M., *Solutiones Contradictionum...super Octavo Physicorum* in *Opera*, vol. IV.

Zotenberg, H., *Catalogues des manuscrits hébreux et samaritains de la Bibliothèque Impériale*, vol. I, i, Paris, 1866.

INDEX OF PASSAGES QUOTED IN THE NOTES

INDEX OF NAMES AND SUBJECTS

Synthese Historical Library

Texts and Studies in the History of Logic and Philosophy

Series Editor: Norman Kretzmann (*Cornell University*)

1. M.T. Beonio-Brocchieri Fumagalli: *The Logic of Abelard*. Translated from Italian by S. Pleasance. 1969 ISBN 90-277-0068-0
2. G. W. Leibniz: *Philosophical Papers and Letters*. A Selection, translated and edited, with an Introduction, by L. E. Loemker. 2nd ed., 2nd printing. 1976 ISBN 90-277-0008-8
3. E. Mally: *Logische Schriften*. Grosses Logikfragment – Grundgesetze des Sollens. Herausgegeben von K. Wolf und P. Weingartner. 1971 ISBN 90-277-0174-1
4. L. W. Beck (ed.): *Proceedings of the Third International Kant Congress*. 1972 ISBN 90-277-0188-1
5. B. Bolzano: *Theory of Science*. A Selection with an Introduction by J. Berg. Translated from German by B. Terrell. 1973 ISBN 90-277-0248-9
6. J. M. E. Moravcsik (ed.): *Patterns in Plato's Thought*. 1973 ISBN 90-277-0286-1
7. Avicenna: *The Propositional Logic*. A Translation from *Al-Shifā': al-Qiyās*, with Introduction, Commentary and Glossary by N. Shehaby. 1973 ISBN 90-277-0360-4
8. D. P. Henry: *Commentary on* De Grammatico. *The Historical-Logical Dimensions of a Dialogue of St. Anselms's*. 1974 ISBN 90-277-0382-5
9. J. Corcoran (ed.): *Ancient Logic and its Modern Interpretations*. 1974 ISBN 90-277-0395-7
10. E. M. Barth: *The Logic of the Articles in Traditional Philosophy*. A Contribution to the Study of Conceptual Structures. 1974 ISBN 90-277-0350-7
11. J. Hintikka: *Knowledge and the Known*. Historical Perspectives in Epistemology. 1974 ISBN 90-277-0455-4
12. E. J. Ashworth: *Language and Logic in the Post-Medieval Period*. 1974 ISBN 90-277-0464-3
13. Aristotle: *The Nicomachean Ethics*. Translation with Commentaries and Glossary by H. G. Apostle. 1974 ISBN 90-277-0569-0
14. R. M. Dancy: *Sense and Contradiction*. A Study in Aristotle. 1975 ISBN 90-277-0565-8
15. W. R. Knorr: *The Evolution of the Euclidean Elements*. A Study of the Theory of Incommensurable Magnitudes and its Significance for Early Greek Geometry. 1975 ISBN 90-277-0509-7
16. Augustine: *De Dialectica*. Translated with Introduction and Notes by B. D. Jackson from the Text newly edited by J. Pinborg. 1975 ISBN 90-277-0538-9
17. Á. Szabó: *The Beginnings of Greek Mathematics*. Translated from German. 1978 ISBN 90-277-0819-3
18. Juan Luis Vives: *Against the Pseudodialecticians*. A Humanist Attack on Medieval Logic. Texts (in Latin), with Translation, Introduction and Notes by R. Guerlac. 1979 ISBN 90-277-0900-9
19. Peter of Ailly: *Concepts and Insolubles*. An Annotated Translation (from Latin) by P. V. Spade. 1980 ISBN 90-277-1079-1
20. S. Knuuttila (ed.): *Reforging the Great Chain of Being*. Studies of the History of Modal Theories. 1981 ISBN 90-277-1125-9
21. J. V. Buroker: *Space and Incongruence*. The Origin of Kant's Idealism. 1981 ISBN 90-277-1203-4

Synthese Historical Library

22. Marsilius of Inghen: *Treatises on the Properties of Terms*. A First Critical Edition of the *Suppositiones, Ampliationes, Appellationes, Restrictiones* and *Alienationes* with Introduction, Translation, Notes and Appendices by E. P. Bos. 1983 ISBN 90–277–1343–X
23. W. R. de Jong: *The Semantics of John Stuart Mill*. 1982 ISBN 90–277–1408–8
24. René Descartes: *Principles of Philosophy*. Translation with Explanatory Notes by V. R. Miller and R. P. Miller. 1983 ISBN 90–277–1451–7
25. T. Rudavsky (ed.): *Divine Omniscience and Omnipotence in Medieval Philosophy*. Islamic, Jewish and Christian Perspectives. 1985 ISBN 90–277–1750–8
26. William Heytesbury: *On Maxima and Minima*. Chapter V of *Rules for Solving Sophismata*, with an Anonymous 14th-century Discussion. Translation from Latin with an Introduction and Study by J. Longeway. 1984 ISBN 90–277–1868–7
27. Jean Buridan's *Logic. The Treatise on Supposition. The Treatise on Consequences*. Translation from Latin with a Philosophical Introduction by P. King. 1985 ISBN 90–277–1918–7
28. S. Knuuttila and J. Hintikka (eds.): *The Logic of Being*. Historical Studies. 1986 ISBN 90–277–2019–3
29. E. Sosa (ed.): *Essays on the Philosophy of George Berkeley*. 1987 ISBN 90–277–2405–9
30. B. Brundell: *Pierre Gassendi: From Aristotelianism to a New Natural Philosophy*. 1987 ISBN 90–277–2428–8
31. Adam de Wodeham: *Tractatus de indivisibilibus*. A Critical Edition with Introduction, Translation, and Textual Notes by R. Wood. 1988 ISBN 90–277–2424–5
32. N. Kretzmann (ed.): *Meaning and Inference in Medieval Philosophy*. Studies in Memory of J. Pinborg (1937–1982). 1988 ISBN 90–277–2577–2
33. S. Knuuttila (ed.): *Modern Modalities*. Studies of the History of Modal Theories from Medieval Nominalism to Logical Positivism. 1988 ISBN 90–277–2678–7
34. G. F. Scarre: *Logic and Reality in the Philosophy of John Stuart Mill*. 1988 ISBN 90–277–2739–2
35. J. van Rijen: *Aspects of Aristotle's Logic of Modalities*. 1989 ISBN 0–7923–0048–3
36. L. Baudry: *The Quarrel over Future Contingents (Louvain 1465–1475)*. Unpublished Latin Texts collected and translated in French by L. Baudry. Translated from French by R. Guerlac. 1989 ISBN 0–7923–0454–3

THE NEW SYNTHESE HISTORICAL LIBRARY
Texts and Studies in the History of Philosophy

37. S. Payne: *John of the Cross and the Cognitive Value of Mysticism*. An Analysis of Sanjuanist Teaching and its Philosophical Implications for Contemporary Discussions of Mystical Experience. 1990 ISBN 0-7923-0707-0
38. D.D. Merrill: *Augustus De Morgan and the Logic of Relations*. 1990 ISBN 0-7923-0758-5
39. H. T. Goldstein (ed.): *Averroes' Questions in Physics*. 1991 ISBN 0-7923-0997-9

Kluwer Academic Publishers – Dordrecht / Boston / London

GPSR Compliance

The European Union's (EU) General Product Safety Regulation (GPSR) is a set of rules that requires consumer products to be safe and our obligations to ensure this.

If you have any concerns about our products, you can contact us on ProductSafety@springernature.com

In case Publisher is established outside the EU, the EU authorized representative is:

Springer Nature Customer Service Center GmbH
Europaplatz 3
69115 Heidelberg, Germany

Batch number: 10761599

Printed by Printforce, the Netherlands